Meteorological Office Great Britain., Stephen Joseph Perry

Report on the Meteorology of Kerguelen Island

Meteorological Office Great Britain., Stephen Joseph Perry

Report on the Meteorology of Kerguelen Island

ISBN/EAN: 9783744720069

Printed in Europe, USA, Canada, Australia, Japan

Cover: Foto ©berggeist007 / pixelio.de

More available books at **www.hansebooks.com**

Official No. 37.

REPORT

ON THE

METEOROLOGY OF KERGUELEN ISLAND.

BY

REV. S. J. PERRY, S.J., F.R.S.

Published by the Authority of the Meteorological Council,

LONDON:
PRINTED FOR HER MAJESTY'S STATIONERY OFFICE,
AND SOLD BY
J. D. POTTER, 31, POULTRY; AND EDWARD STANFORD, 6, CHARING CROSS.
1879.
Price Three Shillings.

GENERAL INTRODUCTION.

THE observations included in this Report on the Meteorology of Kerguelen Island were taken in the winter of 1840 by Sir James Ross, R.N., in the Antarctic Expedition, in January 1874 by Sir G. Nares, R.N., in H.M.S. "Challenger," and in the summer months of 1874-5 by the transit of Venus observers. It was at first intended to reduce merely the last-named series of observations, which were taken during the months of November and December 1874, and of January and February 1875; but as these gave only the summer season, it was resolved to add the winter observations of May, June, and July 1840. Finally, the observations of January 1874, taken under very different circumstances at various parts of the north and east coasts of the Island, were considered to be a valuable addition to the summer observations at the fixed observatory of Royal Sound, and these therefore have been included in this Report.

In all three series the mean results are given in the form of curves, and thus the relations existing between the various elements of any series may be detected at a glance, and the results of the different series can be readily brought into comparison without the labour of any previous reduction.

In the year 1874 a Paper, containing the results of the discussion of the Marine Observations existing in the Office for the region about Kerguelen Island during the month of December, was read before the Meteorological Society, and published in its Quarterly Journal, Vol. II. (new series). This Paper has been reprinted as an Appendix to the present Report.

S. J. PERRY.

TABLE OF CONTENTS.

1. TRANSIT OF VENUS EXPEDITION.

KERGUELEN SUMMER.

REPORT

ON THE

METEOROLOGY OF KERGUELEN ISLAND.

THE KERGUELEN SUMMER.

From Observations taken during the Government Transit of Venus Expedition of 1874.

THE island of Kerguelen is situated in the central portion of the South Indian Ocean, about midway between South Africa and Australia, and a few hundred miles south of the track of the Australian clippers that round the Cape of Good Hope. Its latitude and longitude are approximately 50° S. and 70° E. of Greenwich, and its extent 90 miles from E. to W. and 45 from N. to S. It has never been permanently inhabited but it was formerly a favourite resort of those engaged in the seal fishery, and it is still frequented by a few American sealers, who employ three schooners and a small barque for the collection of skins and for their transmission to the United States.

The general aspect of the country is dreary in the extreme, and it has hence received the uninviting name of the "Island of Desolation." Not a tree or large shrub is to be seen anywhere, and all that meets the eye in every direction is rock and lake and bog. Near the East coast of the island the hills are mostly flat-topped basaltic elevations, rising rather abruptly to no great height, whilst in the South-east sharp mountain peaks form a marked feature in the landscape. The freshwater lakes are numerous, and the whole country, at least near the eastern extremity, is intersected by an endless succession of inlets, or arms of the sea, which stretch sometimes fully 12 miles inland.

The thick fogs and high winds of the entire region in which Kerguelen is situated are proverbial, and H.M.S. "Volage" and "Supply," which conveyed the astronomers to their desolate station, experienced the full force of the violent storms and enormous waves, which follow each other in rapid succession. We were almost in sight of the island on October 6th, but dared not approach nearer on account of the violence of the wind and the thickness of the mist. After two days of suspense and of great suffering the wind abated, and we steered for the entrance to Royal Sound, situated at the extreme Eastern end of the island.

Though it was then already the middle of spring the land was entirely covered with snow from the tops of the mountains down to the water's edge, and on October 10th and 14th we had exceptionally heavy falls of snow during great part of the day. An excellent station for the astronomical observations was found by the aid of Captain

M 613.　　　　　　　　　　　A

Bailey, of the "Emma Jane," and we at once took up our position there, and began the landing of stores and the erection of huts and instruments. The anchorage for the vessels was all that could be desired, and the supply of fresh water excellent and abundant. The site was well protected by distant hills to the North and West, but not in such a way as to interfere with the horizon. We chose a small plateau where the ground was fairly level, and consisted partly of the bare rock and partly of a soft moist soil covered with azorella, with here and there a Kerguelen cabbage or other small plant. Our first care was to drain the spot, and in this we succeeded so well that the dust on our instruments was one of the chief difficulties we had to contend with during the following months. The exact position of our station, where all the meteorological observations were taken, was 49° 25′ 11″·5 S. and 4h. 39m. 34·3s. E. of Greenwich.

The best idea of the general nature of the ground and of the characteristic features of the landscape may be gathered from the pictures on the frontispiece to this Report.

No. 1 marks the positions of the three British stations E 1, E 2, and E 3, and of the American station A, on the shores of Royal Sound. The distance between E 1 and E 2 is about six miles in a direct line.

No. 2 shows the nature of the hills protecting the meteorological station on the North-west and West; and also the general appearance of the ground throughout the island.

No. 3 represents the German station at Betsy Cove, some 10 miles north of the American post. The marshy ground in front gives a fair notion of those portions of the soil that are not covered with loose stones and the débris of rocks.*

No. 4 presents a view of the hills and lakes with which the island is covered.

No. 5 gives the state of our station E 3, near the entrance to Royal Sound, shortly before our Midsummer day, the snow lying thick on the ground.

During our journey from the Cape of Good Hope to Kerguelen the dry and damp bulb thermometers were daily read by the Sappers attached to the Expedition, and they also took a continuous series of readings of the ship's barometer with other observations; but as these were all made with the main object of preparing the observers for their duties at Kerguelen, and as the results would add little to the log of H.M.S. "Volage," I have not thought it advisable to attempt their reduction.

The Meteorological Office provided us with the full equipment of a meteorological observatory. Barometers and aneroids; dry and damp bulb thermometers, with screen; maximum and minimum thermometers for the shade, also with screen; a solar thermometer; a minimum for the grass; thermometers for earth temperatures, and others for the sea; and finally, a small self-recording anemometer, a rain-gauge, and some hydrometers.

The observations extended over the four summer months of November, December, January, and February, and were taken every two hours, both day and night, the work being done entirely by Corporal Wright of the Royal Engineers, assisted by Sappers Hilbert and Wilson, under the immediate superintendence of the general chief of

---

* The results of the German Expedition are published in *Annalen der Hydrographic und Maritimen Meteorologie*, 1875, p. 106.

the astronomical expedition. After a careful study of the results, the general character of the work appears very satisfactory, but there is here and there some slight evidence of a few repetitions of single readings.

THE ATMOSPHERIC PRESSURE.

The barometer used was a Standard No. A 3 (345) by Casella. It had been compared at Kew in April 1873, the correction for index error and capillarity being $+ 0·015$ at $30·5$ in. and $+ 0·007$ at $28·5$ in. It could not be re-compared on its return, as it was unfortunately found broken. This instrument was suspended near a window in the central room of the dwelling-house. The attached thermometer, which was read every two hours along with the barometer, shows that the temperature inside the house was occasionally very high. I should certainly have preferred a more equable temperature, as some corrections would then have been less serious. The height of the cistern above the sea level was 50 feet. The observations entered in Table I. have all been corrected for temperature and height above sea-level, as well as for index error and capillarity. The mean value for each day will be found in the table, and also a graphical representation of the maxima and minima in the curves traced for each month.

The following figures, which give the differences between the monthly mean and the mean for each hour, show that the daily range is rather irregular, and differs considerably from month to month :—

—	2 a.m.	4.	6.	8.	10.	Noon.	2 p.m.	4.	6.	8.	10.	Midnight.
	in.	in.	in.	in.	in.	iu.	in.	in.	in.	in.	in.	in.
November -	+·016	0	+·008	+·002	+·013	−·004	−·011	−·020	−·017	0	+·006
December -	+·009	+·006	+·016	+·006	+·008	−·020	−·007	−·003	−·012	−·008	+·001	+·005
January -	−·005	−·007	−·007	−·003	−·002	+·003	+·008	−·003	0	+·007	+·010	−·006
February -	+·032	+·009	+·011	−·001	−·016	−·005	−·005	0	+·010	−·012	−·115	−·007
Summer -	\+·013	+·002	+·007	+·001	+·001	−·007	−·004	−·007	−·005	−·002	−·001	−·001

The readings are therefore higher in the morning hours than in the afternoon.

The mean daily range for January is small compared with that of the other months, and the large value for February is due entirely to the high readings at 2 a.m.

Table II. gives the rate of change in the height of the column of mercury for every two hours, and may be of great use in studying the barometrical indications in connexion with the beginning and progress of storms.

The mean rapidity of change varies very regularly from hour to hour as follows :—

2 a.m.	4.	6.	8.	10.	Noon.	2 p.m.	4.	6.	8.	10.	Midnight.
·006	·008	·008	·005	·004	·002	·003	·003	·007	·005	·002	·004

The mercury therefore moves most rapidly from 4 to 6 a.m., and at 6 p.m., and least so at noon and 10 p.m.

The most important barometrical results are grouped in the following form:—

—	November.	December.	January.	February.	Summer.
	ins.	ins.	ins.	ins.	ins.
Mean reading • • •	29·658	29·462	29·406	29·610	29·534
Absolute maximum • •	30·293	30·038	29·821	30·183	30·293
,, minimum • •	28·848	28·391	28·440	28·759	28·391
Extreme range • • •	1·345	1·647	1·381	1·424	1·902
Highest daily mean • •	30·257	29·965	29·727	30·073	30·257
Lowest ,, ,, • •	29·165	28·926	28·936	28·901	28·901
Range • • •	1·092	1·039	0·791	1·172	1·356
Greatest daily range • •	0·831	1·094	0·905	0·874	1·094
Least ,, ,, • •	0·073	0·081	0·121	0·064	0·064
Greatest hourly change • •	0·107	0·126	0·119	0·159	0·159

THE TEMPERATURES.

Dry and Damp Bulb Thermometers.

The thermometers read at all the even hours day and night were dry-bulb Pastorelli No. A 195, and damp-bulb Pastorelli No. A 198.

No. 195 was compared at Kew in August 1873, the corrections being : at 40°, −0°·1 ; 60°, + 0°·1 ; 80°, 0.

On our return a comparison made by Mr. R. Strachan, of the Meteorological Office, in January 1877, gave a correction of −0·4 at 47°. The adopted corrections were therefore : at 30°, −0·4 ; 40°, −0·3 ; 50°, −0·2 ; 60°, −0·1.

No. 198, from comparisons at Kew in August 1873, required the corrections : at 40°, −0·1 ; 60°, 0 ; 80°, 0. During the return journey this instrument was broken.

This pair of thermometers was placed in a wooden screen made by the carpenter of H.M.S. "Volage," on the model of the screen furnished by the Meteorological Office which was carried away during the storm on October 6th, which destroyed most of our live stock, tore off a cutter from the davits, and flooded all our cabins and saloons. The new screen* was fastened on the south side of the porch of the wooden dwelling at 4 feet above the ground. This position was considered the best possible under the circumstances, as the protection from the wind was excellent, and the sun could only reach the spot in the early morning hours. Moreover, the wooden walls of the house were double, and there was a door excluding the inner air from free access to the porch, on the outside of which the thermometer screen was suspended. I am afraid, however, that the wooden hut, when unduly heated by the fire within, or by the solar rays in fine weather, may have had some effect in raising the temperature. A few

---

* The screen was louvred on three sides. Its internal dimensions were, in inches, height 15·5, width 7·5, depth 4·5.

readings of the dry and damp bulb have in consequence been enclosed within brackets, and excluded from the means. In November the damp bulb was sometimes incorrectly read, probably owing to inexperience in the management of the moistened thread. These readings do not enter into the means. The readings of the other months are quite satisfactory.

Previous to the commencement of the regular observations the weather had been very severe. Heavy falls of snow and high winds were frequent, and an island in Royal Sound, the whole of whose rocky south side was in the middle of October still covered with enormous icicles, bore witness to the hardness of the winter's frost. On October 25th an interesting phenomenon was seen by the early risers. The snow had fallen rapidly for some time in very large flakes, and as the sea was at the time perfectly smooth, and at a low temperature, the snow rested on the water to the depth of at least half an inch, and remained there for good part of an hour after it had ceased to fall. The morning breeze cleared all away quite suddenly.

Table III. contains the corrected readings of the dry bulb, and Table IV. those of the damp bulb. The mean daily value and the monthly mean for each hour are given with both tables. A few figures will show at once the leading features of Table III.:—

——	November.	December.	January.	February.	Summer.
Mean temperature from dry bulb -	40°.41	43°.75	44°.19	45°.59	43°.49
Highest reading of „ „ -	57.9	57.9	62.7	63.2	63.2
Lowest „ „ „ „ -	28.6	32.8	34.4	35.0	28.6
. Extreme range - -	29.3	25.1	28.3	28.2	34.6
Highest daily mean of dry bulb -	46.0	50.9	52.2	53.9	53.9
Lowest „ „ „ „ „ -	33.3	37.8	39.4	37.3	33.3

The hourly change for the four summer months, obtained by taking the differences between the mean of all the readings and the means for each separate hour, is the following:—

2 a.m.	4.	6.	8.	10.	Noon.	2 p.m.	4.	6.	8.	10.	Midnight.
−4°.02	−4°.07	−0°.70	+2°.28	+4°.24	+5°.16	+3°.09	+2°.11	+0°.03	−2°.28	−3°.10	−3°.39

The highest reading was therefore about noon, and the lowest between 3 and 4 a.m., and the values at 6 a.m. and 6 p.m., differ little from the mean. Being somewhat afraid that the daily variation might be influenced by the exposure of the screen to the morning sun, or by the inside temperature of the hut, I made a similar calculation to the above, having first excluded all the days on which the morning was fine, and then again confined myself to perfectly cloudy days, but the results only gave a confirmation of the above figures, with the sole exception of the minimum being once somewhat earlier. The given values may therefore be adopted as fairly accurate.

The Humidity.

The mean degree of humidity for each day, calculated from the mean results given in Tables III. and IV., furnish the following table:—

—	1.	2.	3.	4.	5.	6.	7.	8.	9.	10.	11.	12.	13.	14.	15.
November -	—	—	—	—	—	84	86	85	77	92	—	84	94	93	84
December -	65	66	79	70	81	73	80	64	81	73	71	69	76	88	86
January -	80	82	75	87	80	72	96	82	75	63	89	83	74	77	75
February -	84	84	76	76	71	79	81	54	83	95	83	74	72	82	77

(continued.)

—	16.	17.	18.	19.	20.	21.	22.	23.	24.	25.	26.	27.	28.	29.	30.	31.
November -	88	77	90	81	77	71	71	64	79	74	80	62	85	90	75	—
December -	73	80	77	68	88	76	81	79	81	78	87	77	70	77	94	77
January -	77	74	78	94	88	86	74	72	81	71	70	81	71	80	79	88
February -	73	78	86	85	81	89	—	—	—	—	—	—	—	—	—	—

We hence obtain for the several months the following values:—

—	November.	December.	January.	February.	Summer.
Mean humidity -	81	77	79	79	79
Greatest ,, -	94	94	96	95	96
Least ,, -	64	64	63	54	54

The following are the values of the mean humidity for each hour of observation:—

—	2 a.m.	4.	6.	8.	10.	Noon.	2 p.m.	4.	6.	8.	10.	Midt.
November -	85	85	83	82	83	79	79	77	84	87	86	83
December -	83	84	80	77	77	71	74	73	76	81	83	82
January -	86	86	82	80	75	72	72	75	77	82	82	80
February -	86	88	95	74	75	73	74	75	74	91	85	84
Summer -	85·0	85·8	85·0	78·3	77·5	73·8	74·8	75·0	77·8	84·3	84·0	82·3

The humidity is therefore greatest during the night hours, and decreases regularly up to noon, and then increases. It attains its least value shortly before 1 p.m.

The mean daily values of the temperature and humidity are graphically represented in the monthly curves.

Maximum and Minimum in the Shade.

A Meteorological Office wall-screen containing the maximum and minimum thermometers was placed by the side of that protecting the dry and damp bulb thermometers, and was a little more sheltered from sun and wind. This screen was similar to that used for the dry and wet bulb thermometers, but was placed horizontally ; its dimensions were in inches: height, 9·75; width, 17·5; depth, 6. It might perhaps have been preferable to place the screen containing the maximum and minimum in a different position from the dry and damp bulb, as then the agreement of results would afford a stronger proof that the figures give a true record of the actual temperature. Identity of exposure has, however, this great advantage, that each set of instruments is the best possible check on the accurate reading of the other.

The maximum and minimum were both made by Pastorelli, and were compared at Kew in May 1873. The maximum No. A 314 was found to require the corrections: at 40°, + 0·1; 60°, + 0·2; 80°, + 0·3; and the minimum No. A 338 the corrections at 40°, + 0·2; 60°, + 0·2.

Both instruments were read at 9 a.m. and 9 p.m., and the following are some of the more important results, taken from Table V. :—

—	November.	December.	January.	February.	Summer.
Mean temperature from maximum and minimum	42·01	45·18	45·91	46·61	44·94
Highest maximum - - - -	61·4	65·2	64·7	68·2	68·2
Lowest minimum - - - -	28·5	32·3	34·2	30·2	28·5
Extreme range - - - -	32·9	32·9	30·5	38·0	39·7
Highest daily mean - - - -	49·0	56·5	·52·6	55·0	56·5
Lowest „ „ - - - -	33·7	40·1	41·3	38·0	33·7
Mean daily range - - - -	15·28	12·76	16·03	15·38	14·86
Greatest daily range - - - -	24·2	23·7	23· 7	28·8	28·8
Least „ „ - - - -	3·6	7·3	7·9	6·1	3·6

The mean temperature obtained from the two-hourly readings of the dry bulb thermometer was 43°·49'; therefore the adopted mean for the Kerguelen summer is 44°·22, which is 10° lower than for the corresponding season in England, and considerably below the adopted mean for the year.

Solar Radiation.

The thermometers supplied were Casella No. 8126 and No. 17,191, whose corrections determined by the Rev. F. Stow in July 1877, were, for No. 8,126. at 90°, − 0·5; 120°, − 1·5 : 140° − 2·5; for No. 17,191, at 90°, + 0·1; 120°, + 0·5; 140°, − 1·5.

The instruments were read daily at 9 a.m, and 9 p.m.

The chief results may be summed up in a short extract from Table VI. :—

—	November.	December.	January.	February.	Summer.
Mean at 9 a.m. - - -	93·44	93·90	95·99	93·93	94·29
„ „ 9 p.m. - - -	102·26	111·89	110·35	108·70	108·30
Highest - - - -	126·2	128·0	122·5	126·9	128·0
Lowest for 24h. - - -	69·5	86·4	86·5	76·0	69·5

Subtracting from these the corresponding shade maxima we obtain—

—	November.	December.	January.	February.	Summer.
Mean radiation from readings at 9 a.m.	47·03	43·17	45·16	41·50	44·19
„ „ „ „ 9 p.m.	52·97	58·73	57·10	55·47	56·07
Maximum radiation	64·8	71·0	72·9	61·5	72·9
Minimum „	34·2	35·4	37·6	35·4	34·2

The direct heating power of the sun was therefore about equally sensible in December and January, and was then considerably in excess of what was observed in the other two months.

Terrestrial Radiation.

The minimum thermometers for the grass, Casella No. 17,966 and No. 17,967, were compared in January 1873 by the Rev. F. Stow, the corrections at 40° being +0·3 and +0·7 respectively. Collecting a few of the leading results from Table VII. we find :—

—	November.	December.	January.	February.	Summer.
Mean value at 9 a.m.	31·48	35·72	34·97	36·94	34·78
„ „ 9 p.m.	34·07	35·58	39·26	37·67	37·15
Highest reading for 24h.	39·4	44·0	40·8	43·3	44·0
Lowest „ „	24·2	28·5	32·1	34·7	24·2

The amount of radiation deduced from these figures by aid of Table V. is as follows :—

—	November.	December.	January.	February.	Summer.
Mean from readings at 9 a.m.	4·06	2·47	3·71	2·60	3·21
„ „ 9 p.m.	2·92	2·40	1·45	3·43	2·55
From highest reading	2·5	2·0	1·2	2·2	2·0
„ lowest „	4·7	6·2	4·1	1·4	4·7

These numbers are very low, a result which is probably due in part to the nature of the ground over which the minimum thermometer was placed, the grass in Kerguelen being thin and growing only in small patches.

Both the solar and terrestrial radiation thermometers were well exposed, being placed in a perfectly open space at a distance from all hills or huts. Owing to the absence of any fixed place for entries in the form used, the numbers on the thermometers were not registered, and it is now impossible to say which of the two instruments in each case was used. The figures are therefore entered in Tables VI. and VII. uncorrected, and the probable error can be at once deduced from the above corrections.

We can form a good idea of the range of temperature during the summer months from the number of days on which the—

—	November.	December.	January.	February.
Shade temperature was above 60° being	2	2	4	4
Grass minimum „ below 32° „	15	9	6	5

Earth Temperatures.

For these observations, the results of which are contained in Table VIII., and which were taken on a principle closely resembling von Lamont's, a brass tube of width just sufficient to allow of the drawing out of the thermometers, was sunk in the ground on the side of a hill where the earth had been well drained. A lath was made to slide in this tube, and four thermometers were fastened to the lath at the respective distances of one, two, three, and four feet from the top of the tube, which was level with the ground. In reading off these thermometers the lath was drawn up quickly, and the readings taken with all the speed consistent with accuracy; but it is still very probable that readings might sometimes be in excess of the true value from slowness of observation.

The corrections of the four instruments determined at Kew in September 1873, were the following :—

—	40°.	60°.	80°.
Hicks No. A 167 at 1 foot -	0	0	0
„ „ 168 „ 2 feet -	0	0	−0·3
„ „ 170 „ 3 „ -	−0·1	−0·2	0
„ „ 171 „ 4 „ -	0	0	−0·1

I feel confident that the above was the order of the instruments, but the fact is not entered in the log.

It may here be useful to bring the extremes into juxtaposition with the mean values entered for each month :—

—	1 foot.	2 feet.	3 feet.	4 feet.	—	1 foot.	2 feet.	3 feet.	4 feet.
Mean, November	41·57	40·60	38·82	38·67	Mean, December	45·22	44·05	42·15	40·87
Highest „	44·8	43·0	40·9	40·0	Highest „	47·5	45·2	44·0	42·1
Lowest „	38·2	38·5	37·9	37·8	Lowest „	42·2	41·8	40·5	38·9
Range „	6·6	4·5	3·0	2·2	Range „	5·3	3·4	3·5	3·2
Mean, January -	46·01	45·13	43·36	42·41	Mean, February	46·42	45·70	43·69	43·04
Highest „ -	49·5	46·7	44·7	43·5	Highest „	51·8	47·5	44·6	43·9
Lowest „ -	43·1	44·0	42·7	41·8	Lowest „	42·3	44·1	42·7	42·1
Range „ -	6·4	2·7	2·0	1·7	Range „	9·5	3·4	1·9	1·8

—	1 foot.	2 feet.	3 feet.	4 feet.
Mean for Summer -	44·81	43·87	42·0	41·25
Highest „ -	51·8	47·5	44·7	43·9
Lowest „ -	38·2	38·5	37·9	37·8
Range „ -	13·6	9·0	6·8	6·1

From these figures it is evident that the temperature of the soil increases gradually throughout the summer, and that the range diminishes rapidly as we penetrate only slightly beneath the surface.

The thermometers at 1 foot and at 2 feet read almost alike when the temperature is low, but a rapid fall sometimes reaches the first a considerable time before it affects the second, and this would naturally be the case more frequently at 9 p.m. than at 9 a.m., on account of sudden changes of temperature about sunset.

The Wind.
Force and Direction.

The observations entered in Table IX. were made without the use of any instrument. We were provided with a small cup-and-dial anemometer, which was placed on the top of a hill 110 feet above the sea level. The exposure of the instrument was good, and it was read daily at 9 a.m. and 9 p.m., but as the force of the wind sometimes injured or loosened parts of the instrument, and twice blew off one of the cups, the results are not sufficiently trustworthy to be taken into account, even as a check on the estimated force. The hill sheltering the dwelling was W. by S. of the observatory grounds, but being small it could scarcely have affected, even slightly, the estimated results of those winds which blew from W. to S.W. Distant hills to the N. and one rather near us between S.W. and S., may have had some influence, but not so as to vitiate the general results.

The daily range of the force for the summer, represented in miles per hour, gives, at—

2 a.m.	4.	6.	8.	10.	Noon.	2 p.m.	4.	6.	8.	10.	Midnight.
24	22·5	22	23	24	26	25	23·5	23	21	21	22·5

showing a regular double progression in the 24 hours, the chief maximum occurring at noon, and the minimum at 9 p.m.

Number of Times each Force of Wind was observed.

Beaufort scale	0	1	2	3	4	5	6	7	8	9	10	11
In miles per hour	3	8	13	18	23	28	34	40	48	56	65	75
Nov.(3000 observations)	2	106	91	37	30	24	7	3	0	0	0	0
Dec. (372 ,,)	3	45	112	54	57	26	24	16	12	10	11	2
Jan. (372 ,,)	0	20	32	97	63	58	39	32	21	7	2	1
Feb. (252 ,,)	0	5	30	26	48	39	34	39	24	5	2	0

Mild winds prevailed in November, with total absence of severe storms. The strongest winds in this month were from the S.W., which blew twice at the rate of 40 miles an hour. The highest daily average rate for the month was less than 30 miles.

In December the winds were strongest and the storms fiercest. They began only on the 16th. Gentle winds were still, however, very prevalent. The mean daily velocity on the 19th was nearly 50 miles an hour, and it remained above 40 miles per hour throughout the 26th and 27th.

If we except the 7th and 8th, when the mean velocity of the wind remained throughout the 48 hours at 40 miles an hour, the storms in January were less violent and less frequent than in the previous month; but the gentle breezes had also disappeared.

In February strong winds were the general rule, the daily mean velocity being over 40 miles on three separate occasions, and nearly equal to 40 during another storm. The velocity was also more constant than during the other months.

A correct idea of the effect of the wind from each point of the compass may be readily obtained from the following tabular form :—

—	N.	NNE.	NE.	ENE.	E.	ESE.	SE.	SSE.	S.	SSW.	SW.	WSW.	W.	WNW.	NW.	NNW.	Sum and Mean.
NOVEMBER.																	
Number of times observed	1	—	8	2	5	4	20	3	11	13	98	9	76	13	31	6	300
Total number of miles	26	—	218	32	160	94	440	98	256	430	2,828	254	2,380	348	1,206	176	8,846
Mean rate per hour	13	—	14	8	16	12	11	16	12	17	14	14	16	13	18	15	14'5
DECEMBER.																	
Number of times observed	5	2	18	—	8	3	6	—	8	4	29	2	98	14	151	19	567
Total number of miles	356	192	438	—	198	78	156	—	238	124	1,346	94	3,432	554	8,062	1,188	16,456
Mean rate per hour	36	48	12	—	22	13	13	—	15	16	23	24	18	20	27	;32	22'5
JANUARY.																	
Number of times observed	3	—	2	—	—	—	7	—	3	17	5	114	60	153	5		372
Total number of miles	98	—	180	—	—	—	112	—	88	164	754	212	5,414	2,340	8,714	304	18,440
Mean rate per hour	16	—	45	—	—	—	8	—	15	27	22	21	24	20	18	36	25
FEBRUARY.																	
Number of times observed	3	—	1	—	2	1	6	—	—	2	14	88	13	112	1		252
Total number of miles	138	—	36	—	42	36	216	—	—	224	256	804	5,180	734	7,008	80	14,744
Mean rate per hour	21	—	18	—	11	18	18	—	—	56	32	29	29	28	30	40	29
SUMMER.																	
Total number of hours	24	4	58	4	30	16	78	6	44	44	296	60	752	200	904	62	2,582
Total number of miles	608	192	872	32	400	208	924	98	582	942	5,184	1,364	16,406	3,976	24,890	1,808	58,486
Mean rate per hour	25	48	15	8	13	13	12	16	13	21	18	23	22	20	28	29	23

It will be seen at once that the average velocity of the wind increases each month, being respectively 14·5, 22·5, 25, and 29 miles per hour, the rate for the whole summer being 23 miles.

The leading features of the several months are still more clearly brought out by aid of diagrams.

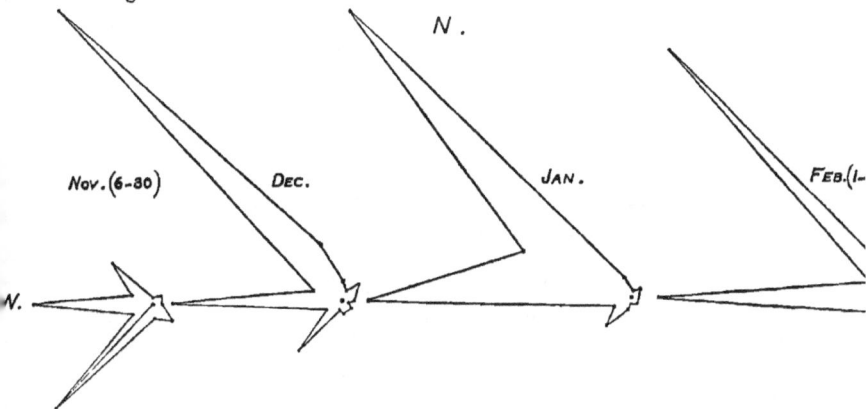

N.

Nov.(6-30)   Dec.   Jan.   Feb.(1-

N.

The absence of easterly winds is more and more strongly marked as the summer advances. The S.W. wind which predominates in November diminishes gradually until it almost disappears in February. On the contrary the frequency as well as the strength of the W. winds is always on the increase; the frequency per cent. for the several months being respectively 25, 29, 31, and 35. For the N.W. wind the corresponding per-centage is 10, 41, 41, and 46, with a gradual increase in the force throughout the summer. The average strength of the N.W. is always greater than that of the W. wind, especially in December.

Taking only eight points of the compass, we have the following results for the Kerguelen summer:—

——	N.	NE.	E.	SE.	S.	SW.	W.	NW.
Total number of miles -	1,608	984	520	1,077	1,102	6,337	19,076	27,782

Or reducing to only four points, we obtain—

——	N.	E.	S.	W.
Total number of miles - - - -	15,489	1,538	4,676	36,783
And therefore per-centage - - - -	26·48	2·63	7·99	62·90

THE RAINFALL.

The rain gauge supplied by the Meteorological Office was exposed on October 31st in a perfectly open position, and was read daily at 9 a.m. and 9 p.m. until February 21st. Table X. furnishes the following results:—

——	November.	December.	January.	February.	Summer.
Number of days of observation - - -	25	31	31	21	108
Number of days on which rain fell - -	8	20	16	11	55
	Ins.	ins.	ins.	ins.	ins.
Amount of rain - - - -	1·95	3·78	3·30	2·30	11·33
Greatest daily fall - - - -	0·75	0·76	0·57	0·47	0·76

Taking into account the number of days in each month on which observations were made, we find that the monthly fall would be in the per-centage proportion of 18·6 in November, 29·3 in December, 25·6 in January, and 26·3 in February. In February the rain fell for 20 hours in succession, but not very heavily.

THE CLOUDS, AND GENERAL WEATHER.

The form and amount of cloud and the general state of the weather were observed at all the even hours of the day and night. No distinction was made between upper and lower clouds. The amount was estimated according to the scale 0, " clear sky "; 10, completely " overcast."

Tho distribution of the clouds throughout tbe 24 hours was fairly regular, though they accumulated somewhat more thickly (7·68) before 10 a.m., and were least numerous (7·23, 7·15) at 10 p.m. and at midnight.

The daily mean amount of cloud for each month was almost constant for tho first three months, but diminished considerably in February.

—	November.	December.	January.	February.	Summer.
Mean amount of cloud - - - -	7·7	7·6	7·7	6·8	7·45
Maximum for 24h. - - - -	9·4	9·8	10·0	9·7	10·0
Minimum for 24h. - - - -	3·8	4·8	3·9	1·5	1·5

January presents the longest duration of unbroken clouds, viz., 42 hours on tho 7th and 8th, and February tho greatest amount of continuous clear sky, tho blue being in the ascendant for 48 hours on the 7th, 8th, and 9th.

From Table XI. wo can determine the relative frequency of tho different kinds of clouds :—

—	Nim.	Cum.	Cum.-s.	Str.	Cir.-c.	Cir.-s.	Cir.	Sml.-c.	Roll.-c.
November - -	39	18	16	3	8	6	4	2	4
December - -	33	34	15	1	6	3	4	1	3
January - -	30	54	10	0	4	0	1	0	1
February - -	30	40	14	9	3	0	3	1	0

The great preponderance of nimbus and cumulus is thus made evident, especially in the month of January, where tho cirrus class is only 1 in 20.

Table XI. furnishes in a similar manner the proportional recurrence of each amount of cloud from 0 to 10 :—

—	0.	1.	2.	3.	4.	5.	6.	7.	8.	9.	10.
November - -	0	1	4	4	5	9	6	8	14	13	36
December - -	0	2	2	5	6	6	10	8	21	12	28
January - - -	0	1	2	5	9	6	7	6	17	15	32
February - -	0	8	6	6	14	8	5	6	10	8	29

Though wo find in each month that a sky completely covered with clouds vastly predominates, and that tho per-centage of such skies is nearly constant, being about 31, still there is a much more even distribution of bluo sky and cloud in tho month of February than in those that precede it. This fact is made still more apparent when we examine tho symbols of each state of weather contained in Tablo X. These symbols are used in accordance with the ordinary Beaufort notation,* and from tho

* Admiral Beaufort's Notation :—Letters to indicate tho State of tho Weather.—b, Bluo sky; c, Clouds (detached) ; d, Drizzling rain ; f, Foggy ; g, Gloomy ; h, Hail ; l, Lightning ; m, Misty (hazy); o, Overcast ; p, Passing showers; q, Squally ; r, Rain ; s, Snow; t, Thunder; u, Ugly (threatening appearance of weather) ; v, Visibility : objects at a distance unusually visible ; w, Dew.

tables we can obtain figures which express the per-centage of each different state of weather observed. Where b,c are taken together a fair proportion of each is supposed to be present in the sky, and it is doubtful which symbol is best suited, neither clearly preponderating. Having examined all the cases, I find that the mean amount of cloud corresponding to the note b,c lies between 5·2 and 5·3.

	b.	b.c.	c.	d.	f.	g.	h.	l.	m.	o.	p.	q.	r.	s.	t.	u.	v.	10.
November -	10·4	6·5	49·0	2·0	0	1·0	1·0	0	2·3	9·7	0·6	1·3	6·8	7·5	0	1·3	0·6	0
December -	10·2	20·1	39·6	2·2	0	1·1	0·3	0	0	12·6	0	0·3	8·8	3·2	0	1·6	0	0
January -	10·0	17·2	46·3	0·5	0	0	0·5	0	0·3	6·7	0·5	0·3	15·3	2·4	0	0	0	0
February -	28·1	13·7	31·3	0·8	0	0	0	0	0	8·8	0	0	16·1	0·4	0	0·8	0	0

As these states of weather may occasionally concur, the per-centage will be somewhat affected, but as concurrent states seldom require to be entered, this will not very materially alter the values. There are several results worth noting in the above weather table; but the total absence of fog, and the few cases of mist with the rare occurrence of gloomy and threatening weather, is the most remarkable, as the island was supposed to be buried in mists and fogs at all seasons of the year. The important forecast of Captain Sir G. Nares, R.N., was verified to the fullest extent, as the position under the lee of the range of mountains N. and W. of Royal Sound was found to afford perfect protection against the fogs of the South Indian Ocean. Thunder and lightning appear to be unknown in the island; and snow fell in each of the summer months. A sky entirely free from cloud was observed only once, but it was fairly clear on 22 occasions.

THE AURORA AUSTRALIS.

Considering that a good watch was kept every night, it is remarkable that the auroral light was observed only on three occasions. On November 13th, between midnight and 2 a.m., the auroral rays were very bright and changeable, and extended from the horizon to an altitude of about 45°, and from S. to S.E., true. The brilliancy and height of the rays varied very much, but there was no great variety in the phenomenon, which consisted mainly of bright vertical rays. On November 24th a very bright aurora was seen in the south between 9 and 10 p.m., and another on January 3rd, between 11 p.m. and midnight.

MARINE OBSERVATIONS.

The Temperature of the Sea Surface.

These observations were made at all the even hours from 6 a.m. to 6 p.m., the instrument used being Pastorelli No. A 193. A comparison at Kew in August 1873, gave as the corrections of this instrument: at 40°, 0; 60°, −0·1; 80°, 0; and after its return Mr. R. Strachan found the correction to be −0·4 at 47° in January 1877. The adopted value has, therefore, been −0·2. The method of observation was to dip the thermometer into the shallow water near the landing-place, and as our range of temperature was not great, I should bo inclined to think that there would not be any very considerable difference botween our roadings and those taken in the open bay. The corrected observations are recorded in Table XII., and the following collation of results shows the romarkable regularity of the changes :—

					6 a.m.	8.	10.	Noon.	2 p.m.	4.	6.	Mean.
November	-	-	-		37·2	38·3	39·1	39·6	39·9	39·3	38·5	38·85
December	-	-	-		39·7	40·7	41·5	42·0	42·1	41·7	41·1	41·37
January	-	-	-		40·5	41·6	42·4	42·9	43·1	42·6	42·1	42·18
February	-	-	-		42·7	43·4	43·8	44·1	44·4	44·1	42·8	43·49
Summer	-	-	-		40·03	41·00	41·70	42·15	42·38	41·93	41·13	41·47

The moan for the day temperature of the sea is therefore about 2° F. lowcr than the mean of day and night for the air, and the maximum occurs between noon and 2 p.m. The highost readings for the several months were respectively 48°·3, 49°·0, 46°·9, and 47°·9, and the greatest daily means 41°·5, 45°·6, 43°·5, and 46°·5.

The Specific Gravity of the Sea-water.

This was observed at the same time as the temperature. The hydrometers used were : 1°. Casella No. A 568; compared by Mr. R. Strachan in April 1874; correction +0·1 at 0; broken on January 20, 1875.

2°. Casella No. A 814, also compared in April 1874; correction +0·1 at 0; and again on return, March 1877; correction +0·3 at 27; adopted correction +0·2.

These instruments were unfortunately only read to the nearest half degree, and I have therefore deemed it advisable to enter the uncorrected values in Table XII. There is very little variation in the readings, with one striking exception on November 28th, when the results were quite abnormal, consisting in a great diminution in the morning, and then a gradual return to the previous value. This gradual return to the former state shows that there was no accidental error in the reading, but there was neither snow, nor rain, nor drizzle, nor any other phenomenon to account for the change. It may be useful to tabulate the corrected means, 1000 being taken as unit :—

			6 a.m.	8.	10.	Noon.	2 p.m.	4.	6.
November	-	-	1027·6	1027·6	1027·8	1027·9	1027·7	1027·8	1027·9
December	-	-	27·9	27·9	28·0	28·1	28·0	28·0	28·0
January	-	-	28·0	27·9	28·0	28·0	28·0	28·0	28·0
February	-	-	28·2	28·2	28·2	28·2	28·2	28·2	28·2

The State of the Sea.

This was registered at all the even hours of the day and night, and the results, contained in Table XII., present such an excellent confirmation of the correct estimation of the force of the wind, that it may be well to enter here the tabular forms corresponding to those given above for the wind :—

Mean Disturbance of the Sea throughout the Day (Scale 0-9).

	2 a.m.	4.	6.	8.	10.	Noon.	2 p.m.	4.	6.	8.	10.	Midn.	Mean.
November -	2·1	1·9	2·1	2·2	2·2	2·4	2·5	2·4	2·4	2·3	2·0	2·0	2·21
December -	2·3	2·1	2·1	2·4	2·2	2·4	2·4	2·3	2·0	1·8	2·0	2·2	2·19
January -	1·8	1·8	1·4	1·4	1·7	2·0	2·0	2·0	1·9	1·5	1·4	1·7	1·72
February -	1·1	0·9	0·9	1·1	1·1	1·2	1·2	1·2	1·2	1·1	1·1	1·1	1·09
Summer -	1·83	1·68	1·63	1·78	1·80	1·98	2·03	1·98	1·88	1·88	1·68	1·63	1·80
Difference from mean	+0·03	−0·12	−0·17	−0·02	0	+0·18	+0·23	+0·18	+0·08	−0·12	−0·17	−0·05	
Difference from mean for wind	+0·9	−0·6	−1·1	−0·1	+0·9	+2·9	+1·9	+0·4	−0·1	−2·1	−2·1	−0·6	

There is the same double daily progression in sea disturbance as in the force of the wind (p. 10), the wind being a little in advance of the sea.

It is obvious from an inspection of the map of Royal Sound at the beginning of this Report that the sea disturbance in Observatory Bay could be no clue to the state of the open sea, especially as the bay was well protected, even on the side of the open Sound, by large islands.

The Height of the Tide.

A tide gauge was erected near our landing place by the officers of H.M.S. "Volage," and read from on board. Readings were also taken on shore from November 29th until February 21st, the hours of observation being the even hours from 6 a.m. to 6 p.m. These are entered in Table XII., which give a fair idea of the daily range :—

—	December.		January.		February.		Summer.	
	ft.	in.	ft.	in.	ft.	in.	ft.	in.
Mean daily range -	3	10½	3	7¾	3	9	3	9
Maximum daily range -	5	11	5	0	5	10	5	11
Minimum „ „ -	2	0	1	9	2	0	1	9
Extreme range - -	6	4	5	4	6	0	6	4

The highest daily readings generally coincided with the time of the meridian transit of the moon, but there was often a priming of the tide, especially in the month of January, and seldom any tendency to lag. The tide seemed to rise highest two days, and to fall lowest three or four days, after new and full moon. There was one notable exception, in which the extreme tides both preceded and followed the full moon.

The Kerguelen Storms.

Having thus far examined the results obtained by passing successively in review the separate tables of observations, I shall now attempt to combine the results, and I think this will be done most effectively by aid of a graphic representation of the data furnished by the different sets of observations. (Vide Plates I. to IV.)

The Barometric curve is formed by dotting down each maximum and minimum as found in the tables, without taking into account the rate at which the mercury may be rising or falling at any given moment between the extreme positions. The deviation from the mean rate between any maximum and minimum can, if required, be ascertained immediately by a glance at the table of barometer rates for each day.

The Thermometer curve consists of points given by the corrected means of the dry bulb readings on each day; and the points of the Humidity curve have been deduced from the daily mean values of the dry and wet bulb thermometers.

The Rain is represented by vertical lines, the scale being 0·5 in. to the inch.

For the Force of the Wind the same scale is used throughout, but the broken line gives the mean values, whilst the dotted lines show the absolute rate of those gales which attained a velocity greater than 40 miles an hour.

The month of November presents a remarkable exception to the other months, as it was entirely free from all violent storms; it will consequently be of less interest to discuss at length any apparent relations that exist between its various meteorological data. The cases of heavy rainfall all occurred with a falling barometer, but the rapid descent of the mercury on the 24th was unaccompanied by any heavy showers. The highest mean velocity of the wind coincides with a rising barometer, and generally with a low temperature. Usually the wind does not freshen with the falling mercury, but rather with the rise which follows each fall, as on the 10th, 15th, 20th. The actual height of the mercury seems to have less connexion than the rate of its variation with the strength of the wind. The increase in the daily mean temperature from the middle of November is very marked.

The first half of December closely resembled the month of November, but some of the heaviest gales occurred during the latter part of the month. The high winds on the 16th and 17th, on the 19th, and 20th, and on the 24th, 25th, and 26th, all came with a falling barometer, but though the mercury then rose rapidly the force of the storm did not abate until the reading was higher than 29 inches. The rapidity in the fall of the mercury is a good indication of the violence of the wind, as may be seen by comparing Tables II. and IX. on the 19th, 24th, and 26th. From the 25th to the 29th the wind blew steadily from the N.W., and though the barometer minimum was the lowest recorded, the rainfall was not heavy. All the strongest winds of this month were from the N.W., beginning mostly from N. and N.N.W. For several days preceding the succession of storms from the 24th to the 27th the mean daily temperature had been rising, and in every case with a strong wind the temperature was high, and a rise of mean daily temperature preceded each storm. The excess of humidity on the 14th, 20th, and 26th coincided with a falling barometer and high temperature, but the greatest humidity of the month presents a notable exception, especially in regard to temperature. The rainfall accompanies or immediately follows every excess of humidity. The highest barometer readings are always accompanied by a rise of mean temperature, and by a W. or N.W. wind. The lowest temperatures occur with a shifting wind, partly at least from the South, with a steady or rising barometer.

M 613. C

In the month of January the severest storm occurred on the 7th and 8th, the highest wind, lowest mean temperature, and heaviest fall of rain coinciding with the lowest barometer, but, as in the great storm on December 26th and 27th, the violence of the tempest did not diminish until the mercury stood above 29 inches. The greatest humidity preceded the heaviest downpour, and the N.W. changed into a W. wind as the storm abated. The second important storm of the month was on the 21st and 22nd, and here again the N.W. changed into a W. wind towards the close, and the violent wind continued with a rising barometer; but this storm differed from the former one, as the direction of the strongest wind was N.N.W., the temperature was high, and no rain fell. The strong N.E. gale on the 29th, with falling barometer and heavy rain, was a solitary exception, the N.W., W.N.W., or W. being the invariable winds for a hard blow and a steady downpour. The greatest degree of humidity coincides always with a falling, but not necessarily with a low barometer, and with various degrees of temperature. The winds of the 17th, 24th, and 31st show that a strong breeze may follow even a slight fall of the barometer during the subsequent rise, but a sharp fall of the mercury is the best sign of a high wind, as is evident from the storms of the 8th, 9th, 21st, and 29th.

The storms in February were less violent but more frequent than in the two preceding months, and the temperature was invariably low during the highest winds. Generally the N.W. changed into a W. wind towards the end of each storm, and the greatest force was usually towards the close. One storm was very exceptional ; a strong W. wind changing into a still stronger S.S.W. The heaviest winds coincided with a rapid rise of the barometer, but the reading was still low. The storm of the 17th agreed with those of December 26th and 27th, and of January 7th and 8th, in commencing with a falling and continuing with a rising barometer. On the 4th, 14th, 20th, and 21st a falling barometer was accompanied by a strong W. or N.W. gale. The rain came invariably when the mercury was low or falling, the maximum humidity preceding the heavy rainfalls. High temperature, with high barometer and little wind, coincided with the minimum humidity, whilst the maximum which preceded severe storms accompanied a rapidly-falling barometer and high temperature. The lowest value of the humidity is coincident with the maximum daily range of the thermometer. The strong W. wind on the 6th with a high barometer preceding a fall, is rather abnormal, as is also the increased force of wind with quiet barometer following a rapid rise on the 13th.

The results for the four summer months may be summed up in a few general remarks. High winds most frequently begin from the N. or N.N.W., passing through N.W. to W. The storm is usually at its height when the wind is N.W., but occasionally the strongest blow comes towards the end of the storm. At the commencement of a gale the barometer is always falling, and the storm may not abate as the mercury rises, provided it is still low. An increase in the violence of the wind is almost invariably indicated by an increased rapidity in the movement of the mercury, which is usually falling. Strong winds appear to be favoured by both extremes of temperature. Excess of humidity accompanied or presently followed by a downpour always comes with a falling barometer, the heaviest rainfall usually coinciding with the lowest barometer. The mercury stands highest with a rising temperature and a W. or N.W. wind. A steady or rising barometer with a shifting wind seems to favour a fall of temperature.

General Concluding Remarks.

A very cursory examination of the results contained in this paper will suffice to convince us that previous impressions concerning the climate of Kerguelen have been far too unfavourable. No month, it is true, is safe from an occasional snow storm, and a high wind each week is the rule rather than the exception, but the sea mists may be entirely avoided by a careful choice of locality, and the temperature in the shade occasionally reaches 60° Fahr., and seldom falls below 32°. Rain is certainly not excessive, and the bill of health during our stay was remarkably clean, which was probably due in great measure to our excellent supply of fresh water, and the abundant crop of Kerguelen cabbage. It may be well here to put on record what we learnt concerning the climate of the island from the most experienced of the sealers, Captain Fuller of the "Roswell King." "The snow," he says, "rarely rests for any length of time " on the lowlands before the month of June, and strong winds seldom last for more than " 12 hours, except in the winter and early spring. The general surface of the island " is not very boggy."

It is not my purpose to give a detailed account of the produce of Kerguelen, but it may not be inappropriate in conclusion to cast a hasty glance at this subject, which has such an intimate connexion with the climate and meteorological conditions. The total absence of all trees and large shrubs is a very marked feature, but their place is supplied to a certain extent by a large quantity of smaller plants, which often cover considerable tracts of country. The well-known Kerguelen cabbage, the azorella, and a species of tea plant, are the most abundant. Grass is rank, though not very plentiful, and there is a fair assortment of ferns, grasses, mosses, fungi, and lichens. The island can apparently boast only of those animals which have been imported, and of which H.M.S. "Volage" carried a large supply. The rabbits were breeding rapidly during our stay, and some young kids were also born. One of the Crozet Islands, which had been stocked with rabbits in a similar manner, is now overrun by these prolific animals. We may also mention that spiders and snails are indigenous, as are also the most curious of the insect tribe, the wingless flies. Kerguelen affords a shelter, and a vast breeding ground, for birds of many species. Wild ducks are there in abundance; they are of small size, but of excellent flavour. The king, the tufted, and the common, or jackass, penguin also abound, and their eggs are collected in large numbers for food by the sealers. Add to these the more majestic wandering albatross, the sooty albatross, the molymauk or skua, eleven distinct species of petrels, the tern, sheathbill, cormorant, and gull, and we see that variety is not wanting at least in the feathered inhabitants of this solitary speck in the Southern Indian Ocean.

I cannot conclude this portion of my Report on the climate of Kerguelen without repeating my admiration of the very efficient way in which my wishes were carried out by the men of the Royal Engineers attached as photographers to the Transit of Venus Expedition. The meteorological observations formed no necessary part of their work, and yet, without in the least interfering with their routine duties, they willingly undertook the laborious task of making eye-observations every two hours day and night for the space of four months. I should be sorry to think that such an excellent example of true devotion in the cause of science should be allowed to pass unnoticed.

Note.—*The tables referred to in the above Report are all preserved in the Meteorological Office.*

THE KERGUELEN WINTER.

From Observations taken on Board H.M.S. "Erebus" and "Terror" during May, June, and July 1840.

The meteorological observations taken at the Transit of Venus Station in Royal Sound, Kerguelen, are confined to the months of November, December, January, and February, and therefore give results for the summer only. Fortunately another series of observations has been made in Kerguelen, and this extends over the months of May, June, and July, and supplies ample materials for determining the conditions of the winter season. These materials are contained in the logs of H.M.S. "Erebus" and "Terror," and date from May 12th to July 20th, 1840. The care with which the meteorological work was carried on under the immediate superintendence of Sir James Ross and Captain Crozier, R.N., renders the labour of reduction a fruitful and satisfactory task, and the results can in most cases be made directly comparable with the corresponding data of the summer series.

The station chosen by Sir James Ross was Christmas Harbour, situated at the northern extremity of the island, and having an entrance a little North of East, of nearly a mile in width. A small bay on the south side slighly increases the width of the entrance for nearly half the depth of the inlet, when there is a sudden contraction to less than one third of a mile. It then gradually diminishes to the head of the bay, which terminates in a level beach of fine dark sand extending quite across, and of about 400 yards in length. The shores on each side are steep, rise in a succession of terraces and platforms, dipping slightly to the N.W., and are surmounted by a remarkable ridge of basalt 1,000 feet above the harbour. The highest hill, which stands at the centre of the north side of the harbour, attains an elevation of 1,350 feet. Its summit is a very distinctly formed oval-shaped crater, measuring about 100 feet across its largest diameter. The narrow isthmus between the head of Christmas Harbour and the N.W. coast, scarcely a mile in breadth, consists of low ridges with intervening swampy ground and two lakes. The ships were warped up to the head of the harbour, and moored in a situation convenient for ready intercourse with the astronomical and magnetic observatories on shore.

NOTE.—The instruments employed on this expedition and their corrections are fully described in "Contributions to our Knowledge of the Meteorology of the Antarctic Regions," Official No. 18, published by the Meteorological Office in 1873.

Some results derived from the log of H.M.S. "Terror," were also published by Dr. Hann in the *Zeitschrift der oesterreichischen meteorologischen Gesellschaft*, vol. xii. p. 100.

The hourly observations, continued day and night on both vessels without interruption during the whole of their stay at the island, were as follows:—

1. Readings of barometer with attached thermometer;
2. Thermometer readings for air and sea;
3. Estimated force and direction of wind;
4. General state of weather;

These were supplemented by hygrometric observations at 9 a.m., noon, 3 p.m., and 9 p.m., and by the amount of rainfall registered on board the "Erebus." The complete and independent record kept by both vessels supplies a most perfect check on the correctness of the results of either. In the reduction of this double series the same order and method have been followed as for the observations of Royal Sound.

THE ATMOSPHERIC PRESSURE.

The barometer readings corrected for temperature are adopted as entered in the logs. To determine the daily range of the barometer, I have excluded the two incomplete days of arrival and departure, taken the differences of the hourly and monthly means, and then combined the two series of results. We thus obtain the following table, the values being as before in decimals of an inch:—

—	1 a.m.	2.	3.	4.	5.	6.	7.	8.	9.	10.	11.	Noon.
May (18 days) -	−·044	−·052	−·053	−·052	−·046	−·040	−·030	−·023	−·007	+·003	+·014	+·021
June -	+·017	+·021	+·017	+·012	+·017	+·021	+·022	+·013	+·023	+·024	+·016	+·002
July (19 days) -	−·032	−·029	−·027	−·018	−·014	−·012	−·008	−·005	+·005	+·015	+·025	+·025

(continued.)

—	1 p.m.	2.	3.	4.	5.	6.	7.	6.	9.	10.	11.	Midn.	Range.
May (18 days) -	+·026	+·013	+·022	+·031	+·028	+·031	+·037	+·034	+·028	+·023	+·017	+·012	0·090
June -	−·010	−·020	−·015	−·016	−·020	−·022	−·025	−·025	−·016	−·010	−·015	−·016	0·049
July (19 days) -	+·021	+·024	+·033	+·237	+·032	+·019	+·011	−·003	−·016	−·022	−·029	−·036	0·073

The close agreement of the results obtained in this case on board the two vessels renders it unnecessary to give both separately, as the point in question is not of very great importance.

The barometer would appear to be more steady in June than in May or July, and the daily progression is fairly marked in each month. The results for May and July agree tolerably well together, but the time scale is almost reversed in June. Thus the mercury stands at its—

—	May.	June.	July.
Highest at - -	7 p.m.	9.10 a.m.	4 p.m.
And falls lowest at -	3 a.m.	7.8 p.m.	Midnight.

The maximum and minimum in July are each three hours in advance of the corresponding phase in May, and the time between the minimum and the following maximum is 16 hours. The mercury therefore would seem in both these months to take twice as long to rise as it does to fall. As July differs so entirely from the other months we can obtain no reliable daily range for the whole winter.

The following table exhibits the principal results of the hourly observations:—

	H.M.S. " EREBUS."			
	May.	June.	July.	Winter.
	Ins.	Ins.	Ins.	Ins.
Mean height of barometer - -	29·491	29·355	29·575	29·474
Absolute maximum - - -	30·160	30·322	30·331	30·331
„ minimum - - -	28·786	28·414	28·752	28·414
Extreme range - - - -	1·374	1·908	1·579	1·917
Highest daily mean - - -	30·065	30·266	30·299	30·299
Lowest „ „ - - -	28·980	28·737	28·996	28·737
Range of „ „ - - -	1·085	1·529	1·303	1·562
Greatest daily range - - -	0·543	1·003	0·822	1·003
Least „ „ - - -	0·048	0·121	0·120	0·048
Mean „ „ - - -	0·314	0·434	0·469	0·406
Greatest hourly change - - -	0·120	0·170	0·190	0·190

The most rapid changes that occurred were a fall of—

0·562 in. in 4 hours on June 18;		0·239 in. in 2 hours on June 28;
1·003 „ 12 „ „ „;		0·374 „ 3 „ July 10;
0·347 „ 3 „ „ 23;		0·174 „ 1 „ „ 11.

The mercury stood above 30 inches from 5 p.m. on July 3rd to 6 p.m. on the 7th.

	H.M.S. " TERROR."			
	May.	June.	July.	Winter.
	Ins.	Ins.	Ins.	Ins.
Mean height of barometer - -	29·492	29·359	29·579	29·477
Absolute maximum - - -	30·186	30·317	30·322	30·322
„ minimum - - -	28·804	28·458	28·776	28·458
Extreme range - - - -	1·382	1·859	1·546	1·864
Highest daily mean - - -	30·067	30·258	30·299	30·299
Lowest „ „ - - -	28·985	28·747	28·987	28·747
Range of „ „ - - -	1·082	1·511	1·312	1·552
Greatest daily range - - -	0·528	1·080	0·813	1·080
Least „ „ - - -	0·039	0·083	0·085	0·039
Mean „ „ - - -	0·305	0·430	0·457	0·397
Greatest hourly change - - -	0·110	0·200	0·172	0·200

The most rapid changes entered in this log were on—

June 18th, a fall of 0·569 in. in 4 hours;		July 15th, a fall of 0·345 in. in 3 hours;
„ 23rd, „ 0·334 „ 3 „ ;		„ 16th, a rise of 0·304 „ 2 „ .

THE TEMPERATURE.

Thermometer in the Shade.

The hourly readings of the thermometer are taken directly from the logs, and the range from hour to hour is calculated as for the barometer:—

	1 a.m.	2.	3.	4.	5.	6.	7.	8.	9.	10.	11.	Noon.
					H.M.S. "Erebus."							
May	−0·1	−0·4	−0·6	−0·6	−0·6	−0·5	−0·4	−0·4	−0·2	+0·4	+0·7	+0·7
June	0	−0·4	−0·2	−0·2	−0·5	−0·4	−0·4	−0·6	0	−0·1	+0·1	+0·5
July	−0·4	−0·5	−0·7	−0·7	−0·6	−0·3	0	−0·1	−0·1	0	+0·8	+0·7
Winter	−0·2	−0·4	−0·5	−0·5	−0·6	−0·4	−0·3	−0·4	−0·1	+0·1	+0·5	+0·6

(*continued.*)

	1 p.m.	2.	3.	4.	5.	6.	7.	8.	9.	10.	11.	Midn.	Range.
					H.M.S. "Erebus."								
May	+1·0	+0·8	+0·6	0	0	−0·2	−0·1	−0·1	−0·1	−0·1	0	−0·2	1·6
June	+0·7	+0·7	+0·8	+0·3	+0·3	+0·4	+0·5	+0·5	+0·4	−0·2	−0·6	−0·6	1·4
July	+0·8	+0·8	+0·7	+0·3	+0·2	−0·2	+0·1	+0·3	0	0	−0·1	−0·2	1·5
Winter	+0·8	+0·8	+0·7	+0·2	+0·2	0	+0·2	+0·2	+0·1	−0·1	−0·2	−0·3	1·4

	1 a.m.	2.	3.	4.	5.	6.	7.	8.	9.	10.	11.	Noon.
					H.M.S. "Terror."							
May	−0·3	−0·6	−0·7	−0·9	−0·7	−0·6	−0·6	−0·6	−0·2	+0·3	+0·7	+1·0
June	−0·2	−0·3	−0·2	−0·2	−0·3	−0·8	−0·8	−0·6	−0·2	−0·1	+0·2	+0·9
July	−0·2	−0·4	−0·3	−0·6	−0·4	−0·6	−0·4	−0·1	−0·1	0	+0·7	+0·9
Winter	−0·2	−0·4	−0·4	−0·6	−0·5	−0·6	−0·6	−0·4	−0·2	+0·1	+0·5	+0·7

(*continued.*)

	1 p.m.	2.	3.	4.	5.	6.	7.	8.	9.	10.	11.	Midn.	Range.
					H.M.S. "Terror."								
May	+0·8	+1·0	+0·5	+0·2	−0·3	−0·1	0	−0·1	−0·1	0	+0·2	+0·2	1·9
June	+0·7	+0·7	+0·7	+0·4	+0·3	+0·3	+0·5	+0·3	+0·3	0	−0·1	−0·3	1·5
July	+0·8	+0·6	+0·3	−0·2	0	−0·2	−0·2	−0·1	+0·1	0	+0·2	+0·1	1·5
Winter	+0·8	+0·8	+0·5	+0·1	0	0	+0·1	0	+0·1	0	+0·1	0	1·4

In both series the minimum occurs between 4 and 6 a.m., and the maximum at 1, 2 p.m.; the deviations from the mean being identical for the two vessels.

During the greater part of the day the progression is fairly regular, but during the hours that immediately precede midnight, there is very little variation of temperature. The reading at 9, 10 a.m. or p.m. differs very slightly from the mean in any case.

The results of greatest interest are best presented in a tabular form :—

	H.M.S. "Erebus."				H.M.S. "Terror."			
	May.	June.	July.	Winter.	May.	June.	July.	Winter.
Mean temperature of the air - - -	35·3	34·8	34·4	34·8	36·6	36·2	36·2	36·3
Highest hourly reading - - - -	42·0	45·0	44·0	45·0	43·0	45·5	45·0	45·5
Lowest ,, ,, - - - -	29·0	29·0	27·0	27·0	31·0	29·5	28·0	28·0
Extreme range observed - - -	13·0	16·0	17·0	18·0	12·0	16·0	17·0	17·5
Highest daily mean - - - -	38·2	39·6	39·0	39·6	39·7	40·2	40·5	48·5
Lowest ,, ,, - - - -	33·0	32·0	27·6	27·6	34·4	33·5	29·2	29·2
Range of ,, ,, - - - -	5·2	7·6	11·4	12·0	5·3	6·7	11·3	11·3
Greatest daily range - - - -	7·5	11·0	11·0	11·0	6·5	12·0	9·0	12·0
Least ,, ,, - - - -	2·0	2·0	2·5	2·0	2·0	2·5	2·5	2·0
Mean ,, ,, - - - -	4·4	6·3	6·8	5·8	4·2	5·9	5·8	5·3
Number of times at or below 32° Fah. - -	40	116	128	284	6	40	44	90
Total number of observations - - -	464	720	464	1,648	445	720	464	1,629

The thermometers were read only to within 0°·5, and the results show that the mean difference between the readings of the two instruments was 1°·5. This difference, due either to the thermometers or to local circumstances, has a very marked effect on the number of times the mercury reached the freezing point. The frost increased in frequency, and also somewhat in intensity from May to July. On board the "Erebus" the thermometer remained below freezing point from 6 p.m. on July 1st until 5 a.m. on the 4th.

The Humidity.

The logs of both ships contain daily determinations of the temperature of the dew point, the observations having been taken on board the "Erebus" at 9 a.m., noon, 3 p.m., and 9 p.m., and only at 9 a.m., 3 p.m. and 9 p.m. on the "Terror." The humidity has been calculated from this data for each observation, and the arithmetical means for each day are contained in the adjoined table :—

	1.	2.	3.	4.	5.	6.	7.	8.	9.	10.	11.	12.	13.	14.	15.
H.M.S. "Erebus."															
May - - -	—	—	—	—	—	—	—	—	—	—	—	—	—	85	86
June - - -	79	68	87	83	78	73	73	83	91	86	88	78	84	78	89
July - - -	88	80	66	83	90	86	96	91	90	83	89	57	87	88	70
H.M.S. "Terror."															
May - - -	—	—	—	—	—	—	—	—	—	—	—	—	—	59	85
June : - -	75	79	52	72	76	75	61	80	85	84	78	71	61	73	72
July - - -	70	47	43	72	87	87	94	85	80	64	75	74	67	57	65

(continued.)

—	6.	17.	18.	19.	20.	21.	22.	23.	24.	25.	26.	27.	28.	29.	30.	31.
					H.M.S. "Erebus."											
May	96	80	82	87	78	81	69	86	84	—	81	—	85	92	89	82
June	84	80	74	81	58	89	84	77	74	79	82	79	66	83	89	—
July	93	95	89	70	—	—	—	—	—	—	—	—	—	—	—	—
					H.M.S. "Terror."											
May	—	65	68	66	54	51	61	61	80	53	69	50	65	—	69	—
June	89	62	62	86	50	82	87	68	70	71	64	67	43	70	81	—
July	94	80	75	58	—	—	—	—	—	—	—	—	—	—	—	—

The methods of observation on board the two vessels must have been somewhat dissimilar, to give such a marked difference of result. From the observations of the "Terror," Kerguelen would appear to have a very dry climate, which can scarcely be reconciled with the known amount of rain and snow; but from the log of the "Erebus," the humidity in winter would differ little from the mean yearly amount in Lancashire. If we tabulate the results according to the hours of observation the two series can be readily compared :—

—	H.M.S. "Erebus."							H.M.S. "Terror."						
	8 a.m.	9.	10.	Noon.	3 p.m.	9.	Mean.	3 a.m.	8.	9.	3 p.m.	9	10.	Mean.
MAY.														
Number of observations	1	8	—	12	7	10	—	—	—	13	10	10	—	—
Mean humidity	85	87	—	81	77	88	83	—	—	62	58	70	—	63
Greatest „	—	96	—	92	84	94	92	—	—	85	70	92	—	82
Least „	—	76	—	71	66	84	74	—	—	45	44	49	—	46
JUNE.														
Number of observations	—	20	—	20	18	21	—	1	2	20	20	18	1	—
Mean humidity	—	80	—	77	78	80	79	87	73	70	70	74	62	71
Greatest „	—	96	—	92	92	93	—	—	—	97	89	87	—	91
Least „	—	47	—	51	65	60	56	—	—	43	52	55	—	50
JULY.														
Number of observations	—	12	1	12	17	11	—	—	—	9	13	8	—	—
Mean humidity	—	88	80	84	80	88	84	—	—	69	71	72	—	71
Greatest „	—	100	—	100	96	100	99	—	—	94	88	94	—	92
Least „	—	53	—	63	48	78	61	—	—	39	41	47	—	42
WINTER.														
Number of observations	1	40	1	44	42	42	—	1	2	42	43	36	1	—
Mean humidity	85	85	80	81	78	85	82	87	73	67	66	72	62	66
Greatest „	—	97	—	95	91	95	95	—	—	92	82	91	—	88
Least „	—	59	—	62	60	74	64	—	—	42	46	50	—	46

Omitting the single observation at 10 a.m., it appears from the results of the "Erebus," that the humidity diminishes from 9 a.m. to 3 p.m., and then returns by 9 p.m. to its former value. We may also notice the large range in the amount of humidity at a station where rain and cloud are so prevalent. The two series of observations differ less in June than in the other months, but unfortunately they are in general so wide apart that, if we give equal weight to both, it is impossible to deduce any very satisfactory conclusions from the whole.

The Wind.

Force and Direction.

Hourly observations of direction and force are registered as usual in the logs of both vessels, but the observers of the "Terror" have generally abstained from recording repetitions of the same symbol. I have therefore supposed both direction and force to remain unaltered until a change is entered. Combining the mean results of the two series of observations, hourly values of the force of the wind are obtained, which furnish data for the calculation of the daily range and hourly changes:—

—	1.	2.	3.	4.	5.	6.	7.	8.	9.	10.	11.	12.
a.m. -	+2·6	+3·2	+3·0	+2·0	+2·4	+1·9	+0·6	−0·8	−1·5	−1·7	−0·7	−0·1
p.m. -	−0·3	−0·7	−1·3	−2·2	−2·2	−2·8	−2·3	−2·0	−1·5	−0·5	+0·7	+2·4

The wind is therefore strongest during the early morning hours, and below the average from 8 a.m. till 10 p.m. There is a slight increase in force towards the middle of the day, the secondary maximum occurring shortly after mid-day, whilst the principal maximum is at 2 a.m. The wind falls off most at 6 p.m., and also considerably at 10 a.m. The mean daily range of velocity is six miles per hour. There is a very fair agreement in the results of the several months.

Number of Times each Force of Wind was observed.

Beaufort scale -	0	1	2	3	4	5	6	7	8	9	10	11	12	Total Number of Observations.
In miles -	3	8	13	18	23	28	34	40	48	56	65	75	90	
H.M.S. "Erebus."														
May -	16	23	30	88	126	54	54	27	17	5	18	4	0	462
June -	21	103	101	78	98	87	73	80	49	21	8	1	0	720
July -	14	33	36	50	69	119	64	23	21	15	13	7	0	464
H.M.S. "Terror."														
May -	7	5	11	56	72	55	73	73	35	31	19	7	0	444
June -	12	54	67	97	91	71	90	100	109	29	0	0	0	720
July -	9	33	26	13	31	47	82	82	56	30	42	11	1	463

Sir James Ross, in the published account of his voyage, speaks of the almost hurricane violence of the Kerguelen gales, sometimes laying the ships over nearly on their beam ends. On one occasion the whole body of the astronomical observatory was moved nearly a foot, and had not the lower framework fortunately been sunk to a good depth below the level of the ground, it would have doubtless been blown into the sea. The gusts were sudden and Captain Ross was frequently obliged to throw himself down on the beach to prevent being carried into the water. One one occasion a man was driven into the sea by a squall, and nearly drowned. On 45 of the 68 days during which the expedition remained at Kerguelen it blew a gale of wind. The above figures show a falling off in the strongest winds in June, and at the same time a considerable increase in the relative number of gentle breezes. July seems to be the most subject to severe storms. A general idea of the direction and force of the wind is given in the annexed form :—

H.M.S. "EREBUS."

Month	Quantity	N.	NbyE.	NNE.	NEbyN.	NE.	NEbyE.	ENE.	EbyN.	E.	EbS.	ESE.	SEbyE.	SE.	SEbS.	SSE.	StbyE.	Total and Mean.
MAY.	Number of times observed	—	—	—	—	—	—	—	—	—	1	—	8	10	3	—	—	460
	Total number of miles	—	—	—	—	—	—	—	—	—	6	—	149	185	24	—	—	12,052
	Mean rate per hour	—	—	—	—	—	—	—	—	—	8	—	19	19	8	—	—	26
JUNE.	Number of times observed	—	—	—	—	—	—	4	6	4	43	9	1	6	4	—	—	709
	Total number of miles	—	—	—	—	—	—	53	109	42	694	290	8	138	63	—	—	17,941
	Mean rate per hour	—	—	—	—	—	—	13	18	11	16	32	8	23	16	—	—	25
JULY.	Number of times observed	—	—	—	—	—	—	—	—	—	24	7	—	—	1	—	—	459
	Total number of miles	—	—	—	—	—	—	—	—	—	404	203	—	—	8	—	—	12,867
	Mean rate per hour	—	—	—	—	—	—	—	—	—	17	29	—	—	8	—	—	28
WINTER.	Number of hours	—	—	—	—	—	—	4	6	4	68	16	9	16	8	—	—	1,628
	Total number of miles	—	—	—	—	—	—	53	109	42	1,106	493	157	323	94	—	—	42,860
	Mean rate per hour	—	—	—	—	—	—	13	18	11	16	31	17	20	12	—	—	26

(continued.)

Month	Quantity	S.	SbyW.	SSW.	SWbyS.	SW.	SWbyN.	WSW.	WbyS.	W.	WbyN.	WNW.	NWbyW.	NW.	NWbyN.	NNW.	NbyW.	Calm.	Total and Mean.
MAY.	Number of times observed	—	1	—	—	4	—	3	14	96	235	45	13	10	—	—	1	16	460
	Total number of miles	—	8	—	—	72	—	69	256	2,338	6,971	1,280	397	249	—	—	8	48	12,052
	Mean rate per hour	—	8	—	—	18	—	23	18	24	30	28	31	25	—	—	8	3	26
JUNE.	Number of times observed	—	1	—	6	—	—	2	12	58	437	39	23	19	7	—	—	28	709
	Total number of miles	—	23	—	63	—	—	16	175	1,231	12,991	701	818	373	61	—	—	84	17,941
	Mean rate per hour	—	23	—	11	—	—	8	15	21	30	18	36	20	9	—	—	3	25
JULY.	Number of times observed	—	1	—	—	—	—	7	8	16	13	354	16	16	7	—	—	14	459
	Total number of miles	—	13	—	—	—	—	106	94	354	339	10,990	375	61	—	—	—	42	12,867
	Mean rate per hour	—	13	—	—	—	—	15	12	22	26	31	23	9	—	—	—	3	28
WINTER.	Number of hours	—	3	—	6	4	—	7	34	170	1,026	97	53	52	7	—	—	58	1,628
	Total number of miles	—	44	—	63	72	—	106	525	3,913	30,953	2,320	1,600	1,600	61	—	—	174	42,860
	Mean rate per hour	—	15	—	11	18	—	15	15	23	30	24	31	31	9	—	—	3	26

H.M.S. "TERROR."

	N.	NbyN.	NNE.	NEbyN.	NE.	NEbyE.	ENE.	EbyN.	E.	EbyS.	ESE.	SEbyE.	SE.	SEbS.	SSE.	SbyE.
MAY.																
Number of times observed	—	—	—	—	—	—	—	—	—	—	—	23	6	—	—	—
Total number of miles	—	—	—	—	—	—	—	—	—	—	—	454	93	—	—	—
Mean rate per hour	—	—	—	—	—	—	—	—	—	—	—	20	16	—	—	—
JUNE.																
Number of times observed	—	5	—	—	—	—	—	29	10	27	—	7	—	—	3	1
Total number of miles	—	40	—	—	—	—	—	507	310	515	—	203	—	—	84	18
Mean rate per hour	—	8	—	—	—	—	—	17	31	19	—	29	—	—	28	18
JULY.																
Number of times observed	—	—	—	—	—	2	—	—	—	21	—	6	—	1	—	—
Total number of miles	—	—	—	—	—	26	—	—	—	621	—	48	—	23	—	—
Mean rate per hour	—	—	—	—	—	13	—	—	—	30	—	8	—	23	—	—
WINTER.																
Number of hours	—	5	—	—	—	2	—	29	10	48	—	36	6	1	3	1
Total number of miles	—	40	—	—	—	26	—	507	310	1,136	—	705	93	23	84	18
Mean rate per hour	—	8	—	—	—	13	—	17	31	24	—	20	16	23	28	18

(continued.)

	S.	SbyW.	SSW.	SWbyS.	SW.	SWbyW.	WSW.	WbyS.	W.	WbyN.	WNW.	NWbyW.	NW.	NWbyN.	NNW.	NbyW.	Calm.	Total and Mean.
MAY.																		
Number of times observed	—	3	—	3	—	3	18	56	18	296	—	—	—	—	—	—	1	442
Total number of miles	—	34	—	69	—	96	639	1,518	921	10,240	—	—	—	—	—	—	3	14,955
Mean rate per hour	—	11	—	23	—	31	36	27	51	35	—	—	—	—	—	—	3	34
JUNE.																		
Number of times observed	—	—	—	—	—	17	—	46	23	523	—	—	—	—	—	—	8	720
Total number of miles	—	—	—	—	—	506	—	938	789	16,382	—	—	—	—	—	—	24	20,658
Mean rate per hour	—	—	—	—	—	30	—	20	34	31	—	—	—	—	—	—	3	29
JULY.																		
Number of times observed	—	3	—	—	—	13	—	45	23	327	—	—	—	—	—	—	9	475
Total number of miles	—	34	—	—	—	350	—	1,413	1,038	13,035	—	—	—	—	—	—	27	16,943
Mean rate per hour	—	11	—	—	—	27	—	31	45	40	—	—	—	—	—	—	3	37
WINTER.																		
Number of hours	—	3	—	3	—	33	18	147	64	1,146	—	—	—	—	—	—	18	1,637
Total number of miles	—	34	—	69	—	952	639	3,869	2,748	39,657	—	—	—	—	—	—	54	52,556
Mean rate per hour	—	11	—	23	—	29	36	26	43	35	—	—	—	—	—	—	3	32

H.M.S. "Terror" was either more exposed to the wind than the "Erebus," or its observers estimated its force at a somewhat higher figure on the Beaufort scale. The great preponderance of the W. by N. (true) wind renders it useless to attempt to exhibit the results in the simple form of a diagram. In July this prevailing wind was stronger and much more continuous than in the two preceding months. An enumeration of the frequent occasions on which the wind continued unaltered in direction for at least 24 hours may be found practically useful. In every case the direction was W. by N. and the figures give the length of time the wind continued, and the total number of miles:—

H.M.S. "EREBUS."

May - Hours, 30, 29, 26;	Miles, 867, 703, 493.
June - „ 73, 54, 50, 27, 24;	„ 2,714, 1,871, 1,616, 995, 562.
July - „ 52, 50, 46, 36, 36, 32, 25;	„ 1,783, 1,838, 1,983, 1,140, 919, 892, 595.

H.M.S. "TERROR."

May - Hours, 55, 32, 31;	Miles, 1,136, 1,262, 1,262.
June - „ 128, 94, 64, 58, 25;	„ 4,244, 3,247, 2,758, 1,672, 733.
July - „ 66, 59, 48, 29, 24;	„ 3,090, 2,424, 2,220, 921, 911.

If we reduce the number of points of the compass first to 16, and then to 8 and 4, for the sake of direct comparison with the summer series, we obtain for the Kerguelen winter the subjoined results:—

N.	NNE.	NE.	ENE.	E.	ESE.	SE.	SSE.	S.	SSW.	SW.	WSW.	W.	WNW.	NW.	NNW.
						H.M.S. "EREBUS."									
4	—	—	107	649	1,124	449	47	22	54	366	2,325	19,752	17,399	353	35
8	—	26	—	1,529	—	821	—	44	—	713	—	38,038	—	1,507	—
69	—	—	—	2,120	—	—	—	399	—	—	—	40,098	—	—	—
						H.M.S. "TERROR."									
20	20	—	253	821	610	457	104	26	52	511	3,049	24,511	20,827	502	92
40	—	—	—	1,332	—	863	—	94	—	1,341	—	46,933	—	1,252	—
268	—	—	—	2,730	—	—	—	275	—	—	—	49,229	—	—	—

Not only is the amount of wind greater, but the ratio of the westerly wind to that from other points of the compass is higher on board the "Terror," still the difference is not larger than might reasonably be expected from the independent application of a method that can only give a fair approximation. From both logs it is evident that in the winter months the excess of west wind is so great that wind from any other quarter may be considered as only a trifling exception.

The Rainfall.

The times at which rain fell were registered in the logs of both vessels, but the amount collected is only given by the "Erebus." The quantity is entered for each hour, and the principal results are contained in the following summary :—

—	May.	June.	July.	Winter.
Number of days of observation -	20	30	20	70
„ „ on which rain fell -	18	29	19	66
	ins.	ins.	ins.	ins.
Amount collected - - -	6·674	8·796	6·913	22·383
Monthly per-centage - -	34·3	30·2	35·5	100
	in.	in.	in.	in.
Greatest daily fall - - -	0·813	0·692	0·685	0·813

In counting the number of days Snow and Drizzle have been entered as Rain ; the amount is never less than 0·02 inch. In the monthly per-centage the number of days of observation is taken into account.

General Weather.

The state of the weather has been entered in the register of both vessels for every hour of the day and night during their stay in Christmas Harbour, and although there are frequent differences, due partly to position, and probably still more to want of uniformity of judgment, yet the general agreement is very satisfactory, especially during the months of May and June.

It will not, I think, be misleading if I attempt to discover from the hourly observations whether any daily range is apparent in the amount of cloud ; but I shall first exhibit in a tabular form the number of days per month on which each state of weather was observed :—

Beaufort Notation*	-	b.	c.	d.	f.	g.	h.	l.	m.	o.	p.	q.	r.	s.	t.	u.	v.	w.
H.M.S. "Erebus."																		
May 20 (days)	-	19	19	9	0	10	10	0	4	18	9	18	10	12	0	0	2	0
June -	-	30	30	4	1	27	11	0	6	28	9	20	19	20	0	0	6	0
July (20 days)	-	19	19	8	2	18	3	1	12	19	8	27	13	14	0	0	4	0
H.M.S. "Terror."																		
May (19 days)	-	15	15	6	0	1	7	0	3	18	3	14	12	12	0	1	4	0
June -	-	22	24	5	0	13	10	0	8	25	7	22	17	20	0	0	12	0
July (20 days)	-	8	15	7	2	3	2	0	11	19	4	19	11	10	0	0	10	0

* See Note, p. 13.

The log of the "Erebus" always gives, along with the symbols b c, the amount of blue sky, expressed in numbers ranging from 0 to 8, whilst the "Terror" affixes the numerals to either b or c. These numbers enable us to form a good idea of the amount of cloud during the 68 days of the Kerguelen winter. In the adjoining table the number of times the blue sky was registered by the "Erebus" with a figure attached is entered under each hour, and also the average amount of sky visible.

—	1 a.m.	2.	3.	4.	5.	6.	7.	8.	9.	10.	11.	Noon.
Number of days -	32	24	23	23	21	21	21	21	24	30	35	35
Amount - -	3·9	3·8	3·9	3·8	3·4	3·8	3·9	4·0	3·8	3·6	3·9	4·1

(*continued*).

—	1 p.m.	2.	3.	4.	5.	6.	7.	8.	9.	10.	11.	Midn.	Mean for the Day.
Number of days	38	34	29	29	29	24	25	31	32	24	25	30	27·5
Amount -	4·1	3·6	3·4	3·6	3·4	4·1	3·8	4·0	4·0	3·8	4·3	4·3	3·85

The sky would therefore seem to be oftenest clear about 1 p.m.; and the earliest hours of the night appear to be far preferable, from the extent of clear sky, and still more from its frequency, to the hours immediately preceding daybreak. The means of the above numbers teach us that for rather more than 40 per cent. of the hourly observations almost half the sky was free from clouds.

The same figures in the log of the "Erebus" will furnish a table of the relative frequency of the various maximum amounts of the blue sky observed on each day:—

—	0.	1.	2.	3.	4.	5.	6.	7.	8.
May - -	1	0	1	1	7	2	7	1	0
June - -	0	0	1	0	5	4	13	7	0
July - -	1	0	2	0	5	3	7	2	0

We thus see that fully half the sky was generally clear for some time at least in every 24 hours. It is certainly remarkable that in the exposed station of Christmas Harbour, and at presumably the worst season of the year, the blue sky should have been visible on almost every day, and so often on so many days; and that in the land of perpetual fogs and mists, fogs should have been registered only on four different days by either vessel, and mists have occurred on less than one third of the days. The incessantly recurring squalls, and the frequent showers of rain, and hail, and snow, are however quite equal to the reputation of this land of desolation. The vivid flashes observed in the N.W. at 7 a.m. on the 18th of July are the only instance of lightning recorded, and thunder was never heard.

The Temperature of the Sea.

If the observations of sea temperature are treated exactly as those of the air, we arrive at results in accordance with the above :—

					H.M.S. "EREBUS."							
	1 a.m.	2.	3.	4.	5.	6.	7.	8.	9.	10.	11.	Noon.
May - -	-0.3	-0.3	-0.3	-0.4	-0.3	-0.2	-0.2	0	-0.1	0	+0.2	+0.2
June - -	0	-0.1	-0.1	-0.1	-0.2	-0.3	-0.4	-0.2	+0.1	+0.1	+0.1	+0.3
July - -	-0.3	-0.1	-0.1	-0.4	-0.3	-0.4	-0.5	0	-0.2	+0.3	+0.4	+0.4
Winter - -	-0.2	-0.2	-0.2	-0.3	-0.3	-0.3	-0.4	-0.1	-0.1	+0.1	+0.2	+0.3

(*continued.*)

	1 p.m.	2.	3.	4.	5.	6.	7.	8.	9.	10.	11.	Midnt.	Range.
May - -	+0.3	+0.2	0	-0.1	+0.3	0	0	0	0	0	0	-0.1	0.7
June - -	+0.3	+0.2	+0.4	+0.3	+0.2	+0.1	+0.1	0	0	-0.2	-0.4	-0.4	0.8
July - -	+0.6	+0.5	+0.5	+0.2	0	-0.1	-0.1	+0.2	-0.1	-0.3	-0.5	-0.4	0.1
Winter - -	+0.4	+0.3	+0.3	+0.1	+0.2	0	0	+0.1	0	-0.2	-0.3	-0.3	0.8

					H.M.S. "TERROR."							
	1 a.m.	2.	3.	4.	5.	6.	7.	8.	9.	10.	11.	Noon.
May - -	0	-0.1	-0.2	-0.4	-0.3	-0.2	-0.2	-0.3	-0.1	0	+0.3	+0.3
June - -	-0.1	-0.1	-0.1	-0.2	-0.3	-0.3	-0.3	-0.3	-0.1	0	0	0
July - -	0	-0.1	-0.3	-0.4	-0.3	-0.3	-0.1	-0.1	-0.2	0	+0.1	+0.3
Winter - -	0	-0.1	-0.2	-0.3	-0.3	-0.3	-0.2	-0.2	-0.1	0	+0.1	+0.2

(*continued.*)

	1 p.m.	2.	3.	4.	5.	6.	7.	8.	9.	10.	11.	Midnt.	Range.
May - -	+0.5	+0.4	+0.2	-0.1	-0.2	-0.2	-0.2	-0.1	0	-0.1	+0.1	0	0.9
June - -	+0.3	+0.4	+0.4	+0.2	0	0	0	-0.1	0	-0.1	+0.1	0	0.7
July - -	+0.3	+0.3	+0.2	+0.1	0	0	0	+0.1	+0.1	+0.1	+0.1	0	0.7
Winter - -	+0.4	+0.4	+0.3	+0.1	-0.1	-0.1	-0.1	0	0	0	+0.1	0	0.7

The progression from hour to hour is almost as regular as for the air, the minimum occurring from 4 to 7 a.m., and the maximum between 1 and 2 p.m. The temperature preceding midnight is the most uncertain.

M 613.

E

The principal features may be tabulated as before:—

					I.M.S. "ERERUS."				II.M.S. "TERROR."			
					May.	June.	July.	Winter.	May.	June.	July.	Winter.
Mean temperature of the sea	-	-	-	-	37·0	35·8	35·3	36·0	37·0	36·1	35·6	36·2
Highest hourly reading -	-	-	-	-	40·0	40·0	41·0	41·0	40·5	39·5	38·5	40·5
Lowest „ „ -	-	-	-	-	34·5	31·0	31·0	31·0	33·5	32·5	30·0	30·0
Extreme range observed	-	-	-	-	5·5	9·0	10·0	10·0	7·0	7·0	8·5	10·5
Highest daily mean	-	-	-	-	38·0	37·6	36·9	38·0	39·1	38·2	37·4	39·1
Lowest „ „	-	-	-	-	35·9	34·7	33·7	33·7	35·4	34·8	32·8	32·8
Range of „ „	-	-	-	-	2·1	2·9	3·2	4·3	3·7	3·4	4·6	6·3
Greatest daily range	-	-	-	-	3·5	6·0	6·0	6·0	5·5	5·0	6·0	6·0
Least „ „	-	-	-	-	0	1·0	1·0	0	1·0	1·0	1·5	1·0
Mean „ „	-	-	-	-	1·8	3·1	3·1	2·7	2·7	2·6	2·7	2·7

The close agreement between the monthly mean temperatures of the air and sea in the log of the "Terror," and the very slight difference between the temperature of the sea as recorded by the two vessels would naturally lead to the conclusion that the thermometer used for the air on board the "Erebus" required a correction of +1°·5.

THE MONTHLY CURVES.

The curves (Plates V. to X.) are traced in the same manner as in the discussion of the observations taken at Royal Sound. The scales, according to which the curves are drawn, are the following:—For the barometer, 40 millimetres to the inch; for the wind, one millimetre to the mile; for the rain, the absolute depth; and for the thermometer, an arbitrary scale of convenient length.

A study of the synchronous curves shows much more clearly than the preceding figures the stormy nature of a Kerguelen winter. Comparing together the two series of curves, we are struck by the great excess of strong winds registered by II.M.S. "Terror," especially towards the middle of May, and in the month of July, but still we find that in June the "Erebus" occasionally entered in her log a stronger wind than the "Terror."

The only interval of any considerable length during which the barometer could be called steady, viz., from the 18th to the 23rd of May, was also remarkable for the constancy of its mean daily temperature, for the total absence of high winds, and for the diminution of rainfall. The high barometer from May 30th to June 3rd also coincided with a quiet period, and with little rain. But from the 20th to the 22nd of June, and from the 3rd to the 7th of July, although the mercury stood high, the log of the "Terror" registered a fair number of strong winds, and the rain was not much below the average. A daily gale with a rapidly changing barometer was the rule rather than the exception during the whole winter, and the rainfall, though much less in amount when the mercury stood high, was scarcely ever absent for 24 hours.

More steadfast even than the rain was the direction of the wind, which varied only by a few points, except on rather rare occasions. All the long-continued gales came from the W. by N.. and rarely shifted even to W. or W.N.W. The storms on the 26th, 27th, and 29th of May prove that a low barometer is not a necessary condition for the stiffest breeze, and a depression of the mercury is not always the forerunner of a heavy gale, as is shown by the fact that the lowest minimum during the whole period of observation was followed by an almost perfect calm. Still, however, it is generally found that a low barometer and a strong wind are pretty constant neighbours, and it may be considered an almost universal law at Kerguelen that the height of any storm is only reached as the mercury rises after a rapid fall.

The variations of temperature were so slight, that they could have had little influence on the recurrence of storms.

COMPARISON BETWEEN SUMMER AND WINTER.

As this can perhaps be shown most effectively by placing side by side the principal results for the two seasons, I will now present in a tabular form the most important values, referring for any further details to the papers themselves :—

	Summer.	Winter.	
		H.M.S. "Erebus."	H.M.S. "Terror,"
	ins.	ins.	ins.
Mean barometer	29·534	29·474	29·477
Highest reading of barometer	30·293	30·331	30·322
Lowest „ „	28·391	28·414	28·458
Mean temperature	43·5°	34·8°	36·3°
Highest „	63·2	45·0	45·5
Lowest „	28·6	27·0	28·0
Highest daily mean temperature	53·9	39·6	40·5
Lowest „ „	33·3	27·6	29·2
Mean temperature of sea from 6 a.m. to 6 p.m.	41·5	36·1	36·3
Mean humidity	79	82	68
Percentage of N. wind	26·48	0·16	0·52
„ S. „	7·99	0·93	0·53
„ E. „	2·63	4·97	4·02
„ W. „	62·90	93·94	94·93
Percentage of rainy days	51	94	91
	ins.	ins.	
Average fall on „	0·21	0·34	—
Mean amount of cloud (0–10)	7·45	8·05	—

The exposed nature of Christmas Harbour would naturally cause its rainfall to be somewhat in excess of that at Royal Sound, but, even taking the situation into account, it is evident from the above figures that the winter is the rainy season of the year. Allowing for the number of days of observation, the ratio of the quantities of rainfall in winter and summer is as 3 to 1, and the number of rainy days as 13 to 7. The computed humidity, however, is not greatly in excess in the winter, even if we discard altogether the small values obtained from the observations of the "Terror."

The west wind, which was very prevalent in summer, appears to be almost constant during the winter months. This difference in the seasons arises principally from the N.W. winds of Royal Sound being replaced by W. by N. winds at Christmas Harbour.

This may be due to the relative positions of the two stations quite as much as to the season of the year, for Christmas Harbour lies at the extreme north point of the island, whilst Royal Sound is situated at the S.W. side, and is protected by mountain ranges to the W. and N.W.

The difference between the mean temperature in summer and winter is very small, being only about 8° Fahr., and the lowest minimum is nearly identical in the two seasons, but the highest maximum is so much greater in summer than in winter that the extreme range of temperature in one case is almost double what it is in the other. The mean temperature of the sea is 2° Fahr. lower than that of the air in summer, and slightly in excess in winter.

In conclusion, I would remark that although the frequency and violence of the westerly gales. and the almost daily recurrence of rain and snow during the months, fully bears out all that has been related in past accounts of the stormy nature of these inhospitable shores, it is matter for surprise to find so little confirmation of the earlier reports of extreme cold experienced in Kerguelen. To reconcile reports with accurate observations we must necessarily suppose that the former refer only to the weather side of the island, and that this differs very materially from the N.E. coast.

METEOROLOGICAL OBSERVATIONS TAKEN AT KERGUELEN ISLAND

On board H.M.S. "Challenger,"

JANUARY 7TH TO 31ST, 1874.

THE data contained in the log of H.M.S. "Challenger" comprise hourly observations of the barometer with attached thermometer, of the dry and damp bulb thermometer, of the force and direction of the wind, of the upper and lower clouds, and of the general weather, from January the 7th to January the 31st, 1874, together with a few sea temperatures, and remarks on the state of the sea, taken principally when the ship was not in harbour. The form of reduction scarcely differs at all from that adopted in the preceding pages, and thus the results are made more immediately comparable.

THE ATMOSPHERIC PRESSURE.

The instrument used for observing the pressure was a marine barometer, Adie, 418A, compared at the Meteorological Office, London, by R. Strachan, and found to require a correction of -0.001 at 29.5 inches. The height of the barometer cistern above sea-level was $9\frac{1}{2}$ feet.

The hourly readings give the following mean range for the month of January :—

—	1.	2.	3.	4.	5.	6.	7.	8.	9.	10.	11.	12.
A.M.	+ ·001	− ·004	− ·004	− ·012	− ·009	− ·012	− ·015	− ·016	− ·016	− ·021	− ·016	− ·013
P.M.	− ·010	0	+ ·001	+ ·002	+ ·008	+ ·012	+ ·009	+ ·017	+ ·020	+ ·021	+ ·024	+ ·021

These figures agree well with those of the same month in 1875, though they are considerably in excess. In November, December, and February the morning readings were generally higher than those of the afternoon in the summer of 1874–5.

The most important and interesting barometric results are collected in the adjoined tabular form:—

Mean reading for the month	-	-	-	-	29·708 ins.
Absolute maximum	-	-	-	-	30·131
„ minimum	-	-	-	-	29·028
Extreme monthly range	-	-	-	-	1·110
Highest daily mean	-	-	-	-	30·048
Lowest „ „	-	-	-	-	29·304
Range of „ „	-	-	-	-	0·751
Greatest daily range	-	-	-	-	0·726
Least „ „	-	-	-	-	0·084
Mean daily maximum	-	-	-	-	29·857
„ „ minimum	-	-	-	-	29·559
„ „ range	-	-	-	-	0·305

We notice in these values that the daily maxima and minima give identically the same reading for the month as the mean of all the readings.

Comparing the above figures with the corresponding results for 1875 we find the mercury to have stood much higher and to have been more steady in 1874. The only numbers agreeing closely in the two years are those for the range of the daily means.

On four occasions during the 25 days of observation in January 1874 the mercury stood above 30 inches, and it remained above that height twice for the space of 18 hours. The barometer was steadiest from the 23rd to the 25th, when it only varied between 29·5 and 29·4 for 35 hours in succession.

THE TEMPERATURE.

The dry and wet bulb thermometers employed for determining the temperature and hygrometric condition of the atmosphere were compared at Kew in September 1872, the corrections being—

	At 40° Fahr.	60°.	80°.
For the dry bulb thermometer, 91n	−0°·1	−0°·1	−1°·0
„ „ wet „ „ , 93a	−0°·1	0	−1°·0

The hourly variations of the dry and wet bulb do not differ materially in character or in times of maximum and minimum, but only in amount. The differences between the means for each successive hour and the monthly mean furnish the following table:—

—	1.	2.	3.	4.	5.	6.	7.	8.	9.	10.	11.	12.
Dry { a.m.	−2°·1	−2°·2	−2°·3	−2°·1	−2°·2	−1°·2	−0°·8	0	+0°·2	+0°·9	+1°·6	+2°·4
p.m.	+3°·2	+2°·7	+2°·4	+1°·9	+1°·6	+1°·1	+0°·3	−0°·2	−0·8	−1°·1	−1°·4	−1°·7
Wet { a.m.	−1°·5	−1°·6	−1°·6	−1°·6	−1°·5	−0°·7	−0°·2	+0°·3	+0°·5	+0°·7	+1°·1	+1°·5
p.m.	+1°·9	+1°·8	+1°·5	+1°·2	+1°·0	+0°·5	0	−0°·3	−0°·7	−0°·9	−1°·1	−1°·1

There is a single regular progression in both thermometers, the mean reading for the dry bulb occurring at about 8 a.m. and p.m., and for the wet an hour earlier. The dry bulb reaches its maximum shortly after 1 p.m., and its minimum about 3 a.m., whilst the highest reading of the wet bulb occurs between 1 and 2 p.m., and its lowest between 2 and 4 a.m.

The chief results for the two instruments may be briefly tabulated as before:—

	Dry.	Wet.
Mean temperature - - -	44·0°	41·8°
Mean daily maximum - - -	48·4	45·0
„ „ minimum - - -	40·4	38·0
„ „ range - - -	8·0	7·0
Absolute maximum - - -	57·9	55·0
„ minimum - - -	37·9	34·9
Extreme range - - - -	20·0	20·1
Highest daily mean - - -	50·4	49·0 ·
Lowest „ „ - - -	39·2	37·6
Range of „ „ - - -	11·2	11·4
Greatest daily range - - -	17·0	13·2
Least „ „ - - -	1·7	1·8

The mean temperature for January is almost identical in 1874 and 1875, but the extreme range is considerably less in 1874 than in 1875, both on account of the maximum being lower and the minimum higher. The frequent change of station would tend to increase the range, but this was probably more than compensated for by the observations of 1875 being all taken at a land station.

The Humidity.

The humidity has been computed from the mean hourly and the mean daily readings of the dry and moist bulb thermometers. The value found for the month is 82·7. The mean daily amount was on five occasions greater than 90, and six times below 80, and tho extreme values were 96·0 and 66·0. Tho range obtained from tho hourly moans is the following:—

—	1.	2.	3.	4.	5.	6.	7.	8.	9.	10.	11.	12.
A.M.	+5·1	−2·6	+6·1	+4·4	+6·2	+3·7	+4·8	+2·3	−7·3	−0·5	−1·9	−3·9
P.M.	−6·4	−3·6	−4·7	−3·1	−2·6	−3·1	−1·8	−0·2	+0·8	+1·3	+3·0	+5·2

Tho humidity at night is far in excess of that during tho day, tho probable maximum occurring between 3 and 5 a.m., and the minimum about 1 p.m. These results are a good confirmation of tho values obtained for tho summer of 1875. The mean humidity for January was considerably greater in 1874 than in 1875.

THE WIND.
Force and Direction.

Observations were made every hour of both the force and the direction of the wind, 32 points being taken for the direction, and half divisions of tho Beaufort scale for the force. The magnetic declination at Kerguelen makes tho true north equivalent to tho N.E. by N. "magnetic"; this correction has therefore been applied in all cases.

Tho mean hourly variation of the wind's force is a fairly regular simple progression, tho mean values occurring between 7 and 8 a.m. and p.m. and the maximum and minimum at 2 p.m.; and from 3 to 5 a.m. respectively. The following are tho figures, tho force scale varying from 0 to 12, and the quantities entered being the differences between tho monthly mean and the mean for each separate hour:—

—	1.	2.	3.	4.	5.	6.	7.	8.	9.	10.	11.	12.
A.M.	−·3	−·5	−·7	−·6	−·9	−·4	0	+·2	+·2	+·3	+·6	+·5
P.M.	+·7	+·9	+·7	+·5	+·4	+·5	+·1	−·1	−·3	−·4	−·4	−·5

Tho number of times that each force of tho wind was observed will give a good idea of its general strength:—

Beaufort scale	0	0·5	1	1·5	2	2·5	3	3·5	4	4·5	5	5·5
Number of times observed	13	21	105	12	39	80	43	122	18	34	16	31
(continued).												
Beaufort scale	6	6·5	7	7·5	8	8·5	9	9·5	10	10·5	11	12
Number of times observed	13	4	6	2	16	2	9	0	0	0	0	0

Compared with tho results of 1875, the total absence of very severe gales is remarkable, as also tho frequency of mild breezes.

The following table will give a clear notion of the force of the wind from each direction of the compass. The figures express the number of times each wind was observed:—

Beaufort Scale	0	1	1'5	2	2'5	3	3'5	4	4'5	5	5'5	6	6'5	7	7'5	8	8'5	9	Total
N.	—	1	—	—	1	—	—	—	1	—	—	—	—	—	—	—	—	—	3
N. by E.	1	22	—	1	2	1	1	1	—	—	1	—	—	—	—	—	—	—	30
NNE.	—	—	—	—	1	—	1	—	—	—	—	—	—	—	—	—	—	—	2
NE. by N.	—	1	—	—	—	—	—	—	—	—	—	—	—	—	—	—	—	—	1
NE.	—	2	—	—	—	—	—	1	2	—	—	—	—	—	—	—	—	—	5
NE. by E.	—	2	—	—	—	—	—	—	—	—	—	—	—	—	—	—	—	—	2
ENE.	—	—	—	—	—	—	—	—	—	—	—	—	—	—	—	—	—	—	0
E. by N.	—	3	—	—	—	—	—	—	—	—	—	—	—	—	—	—	—	—	3
E.	1	2	—	—	—	—	—	—	—	—	—	—	—	—	—	—	—	—	3
E. by S.	5	5	—	—	—	—	—	—	—	—	—	—	—	—	—	—	—	—	10
ESE.	—	—	—	—	—	—	—	—	—	—	—	—	—	—	—	—	—	—	0
SE. by E.	—	—	—	—	—	—	—	—	—	—	—	—	—	—	—	—	—	—	0
SE.	—	—	—	—	—	—	—	—	—	—	—	—	—	—	—	—	—	—	0
SE. by S.	1	1	—	2	1	2	—	—	—	—	—	—	—	—	—	—	—	—	7
SSE.	—	—	—	—	—	—	—	—	—	—	—	—	—	—	—	—	—	—	0
S. by E.	1	—	—	—	—	—	—	—	—	—	—	—	—	—	—	—	—	—	1
S.	1	—	—	—	—	—	—	—	—	—	—	—	—	—	—	—	—	—	1
S. by W.	3	—	—	2	—	—	—	—	—	—	—	—	—	—	—	—	—	—	5
SSW.	—	—	—	—	—	—	—	—	—	—	—	—	—	—	—	—	—	—	0
SW. by S.	1	5	1	—	5	—	2	—	—	—	—	—	—	—	—	—	—	—	14
SW.	—	—	1	—	—	1	—	—	1	—	4	—	—	—	—	—	—	—	7
SW. by W.	—	7	—	—	3	2	2	—	1	—	—	—	—	—	—	—	—	—	16
WSW.	—	6	5	1	4	—	—	1	—	1	1	1	—	—	—	—	—	—	20
W. by S.	—	3	1	5	5	9	16	2	5	3	—	—	—	—	—	—	—	—	49
W.	—	—	—	1	1	1	12	6	—	—	8	1	1	2	—	1	—	—	34
W. by N.	4	21	1	12	17	8	34	—	6	2	5	7	2	—	—	3	1	—	123
WNW.	—	1	—	1	3	2	12	1	8	1	1	2	—	—	1	1	—	—	34
NW. by W.	1	7	1	6	10	6	12	5	5	7	2	1	—	—	1	—	—	—	64
NW.	—	4	—	3	12	3	9	1	2	—	4	1	—	2	—	2	—	1	44
NW. by N.	2	9	1	5	9	5	20	1	2	3	—	—	—	—	3	—	2		62
NNW.	—	—	—	—	1	1	1	—	—	—	2	—	—	1	1	5	1	6	19
N. by W.	—	2	2	—	6	1	2	—	—	—	2	—	—	1	—	—	—	—	15

We see here at a glance the great prevalence and strength of the winds from the NW. quarter. This table may be conveniently thrown into the form previously adopted :—

—	N.	N. by E.	NNE.	NE. by N.	NE.	NE. by E.	ENE.	E. by N.	E.	E. by S.	ESE.	SE. by E.	SE.	SE. by S.	SSE.	S. by E.
Number of observations	3	30	2	1	5	2	0	3	3	10	0	0	0	7	0	1
Total of miles	49	318	36	8	90	16	0	24	22	67	0	0	0	91	0	6
Rate per hour	16	11	18	8	18	8	0	8	7	7	0	0	0	13	0	6

(continued.)

—	S.	S. by W.	SSW.	SW. by S.	SW.	SW. by W.	WSW.	W. by S.	W.	W. by N.	WNW.	NW. by W.	NW.	NW. by N.	NNW.	N. by W.	Calm.
Number of observations	1	5	0	14	7	16	20	48	34	123	34	64	44	62	19	15	3
Total of miles	5	43	0	174	178	236	303	924	878	2,333	795	1,242	960	1,212	828	280	39
Rate per hour	5	9	0	12	25	15	15	19	26	19	23	19	22	20	44	19	3

A diagram may help to bring out the result in bolder relief. The scale is 4 mm. to each 100 miles.

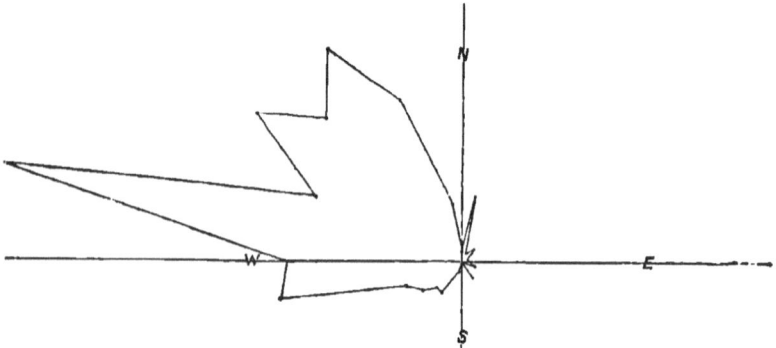

The direction has varied more than in the corresponding month of 1875, as might be expected from the frequent change of station in 1874. The almost total absence of SE. wind is as apparent as before, and the general direction is still NW. by N., though inclined more to W. in 1874. The prevailing wind is here W. by N. and was NW. in 1875.

By reducing the number of points of the compass to 16 we are able more directly to compare the results with those of the other series :—

N.	NNE.	NE.	ENE.	E.	ESE.	SE.	SSE.	S.	SSW.	SW.	WSW.	W.	WNW.	N.W.	NNW.
348	199	102	20	68	33	46	48	30	108	383	883	2507	2582	2186	1574

A further reduction to eight points gives the following :—

N.	NE.	E.	SE.	S.	SW.	W.	NW.
1079	132	113	91	54	739	4684	4225

And if we confine our attention to the four cardinal points the result is—

N.	E.	S.	W.
3255	174	408	7280

which therefore gives the per-centage—

N.	E.	S.	W.
29·3	1·5	3·7	65·5

This does not differ very materially from tho por-contage found for the summer of 1874–5 at Royal Sound, but the E. and S. have a slightly greater influence in the wholo summer than in January alono.

If wo include the 39 milos of wind, for volocitios less than 0·5, wo find a total of 11,157 miles in 587 hours of observation, or a mean velocity of about 19 miles an hour. The longest duration of any wind was that of tho W.N.W., which lasted for 27 consecutive hours.

In the remarks of tho log of H.M.S. "Challenger" we find on the 25th, at 9·45, a.m. "Very heavy squalls from W. by N." are entered, and at 9 p.m. "Wind from W. by S. in tho squalls," although tho general direction was W. by N. On tho 29th, at 1 a.m., the log notos "Squalls very sudden and without warning."

THE CLOUDS AND GENERAL WEATHER.

The upper and lower clouds are entered separatoly in the log, the upper being observed 185 timos during tho 25 days of January and the lower 578 times. The nature of the clouds will bo determined approximately by their position, but it may still be useful to compute tho rolative frequency of each kind of cloud with groator accuracy:—

—	Nim.	Cum.	Cum.-s.	Str.	Cir.-c.	Cir.-s.	Cir.
Lower { alone - - - -	9	326	118	52	1	0	1
{ in combination - -	28	57	5	35	0	0	2
Upper clouds - - - -	1	1	1	47	9	26	99

This table shows cloarly tho very large per-centago of cumulus, and tho comparativo almost absenco of nimbus. The appearauce of tho lattor was extromoly raro except in conjunction with cumulus. Tho prepondoranco of cumulus was not so marked in 1875, but tho por-contago of cumulus and nimbus combined was thon 84.

The mean amount of cloud for the month was 7·1 on tho scale 0–10, and the hourly rango is bost given in a tabular form:—

—	1.	2.	3.	4.	5.	6.	7.	8.	9.	10.	11.	12.
A.M. - -	−·5	−·3	−·1	+·2	+·9	+·6	+·3	+·5	+·4	−·1	−·1	−·3
P.M. - -	−·5	−·5	−·4	−·7	−·1	+·1	+·2	+·4	+·6	+·4	−·1	−·6

There is evidently a double progression, but the timos of maxima and minima aro not marked as clearly as could bo desired. Tho hours immediately following noon and midnight aro those most free from clouds. Tho mean daily amount of cloud was only once in excess of 9·0, and only onco less than 5·0. Tho number of times

APPENDIX.

The Meteorology of December in the Southernmost Part of the South Indian Ocean.—Drawn up from Information received at the Meteorological Office,* by Robert H. Scott, F.R.S.

[FROM THE "QUARTERLY JOURNAL OF THE METEOROLOGICAL SOCIETY" FOR JULY 1874.]

THE meteorology of the part of the South Indian Ocean lying between latitude 45° and 53° S., and longitude 40° and 80° E., in which are the islands of Kerguelen, the Crozets, and Heard's (McDonald's) Island, has recently been discussed in the Meteorological Office with a view of deriving such information respecting the weather in this region in December as might be useful in connexion with any expeditions which may be sent to these islands for observing the transit of Venus in 1874. As it will be many years before more complete data can be procured and prepared, it may be useful to make the present results available to meteorologists and geographers generally.

From the registers kept in this region during the month of December, sets of observations were selected which would give the best daily mean values. They were generally six in number, though sometimes less. A day's observations were never divided, but all entered to the position of the ship at noon. The barometrical observations have been corrected for scale errors, for temperature, and reduced to sea-level. The temperatures have been corrected for any errors of thermometers exceeding half a degree. The specific gravity of the sea is given, corrected for errors of hydrometers and reduced to the standard temperature, 62° F. The directions of the wind have been corrected for variation of the compass.

The ships all sailed eastward through the region; those between 45° and 50° S. averaged latitude 46° 40' S., and those between 50° and 53° S. averaged 51° 12' S.

The years represented are the following :—

1855, 48 days; 1856, 6 days; 1857, 10 days; 1858, 6 days; 1859, 28 days; 1860, 4 days; 1861 6 days; 1870, 10 days: so that the observations embrace 118 days altogether.

For one particular day there are observations from 4 ships; for two other days from 3 ships; and for twenty-two of the days from 2 ships.

The distances apart of the ships were too great for satisfactory synoptic comparisons of weather, but their observations indicate the law of wind in relation to barometric pressure. In applying this law for the purpose of foretelling the wind's direction or its changes in high southern latitudes, the fall of barometer due to change of latitude should be taken into consideration.

The average heights of the barometer show the decrease of pressure for increase of latitude; but there is also considerable difference in respect to longitude. The barometer ranges higher about longitude 50° to 70° E., than to the eastward or westward. The temperature of the air is there also slightly higher. Between latitude 45° and 50° S., from longitude 40° to 80° E., the temperature of the sea increases from 39° to 45°; but between latitude 50° and 53° S., in the same longitudes, the sea is

---

* NOTE.—This paper has been prepared by Mr. R. Strachan, F.M.S.

APPENDIX.

The Meteorology of December in the Southernmost Part of the South Indian Ocean.—Drawn up from Information received at the Meteorological Office,* by Robert H. Scott, F.R.S.

[From the "Quarterly Journal of the Meteorological Society" for July 1874.]

The meteorology of the part of the South Indian Ocean lying between latitude 45° and 53° S., and longitude 40° and 80° E., in which are the islands of Kerguelen, the Crozets, and Heard's (McDonald's) Island, has recently been discussed in the Meteorological Office with a view of deriving such information respecting the weather in this region in December as might be useful in connexion with any expeditions which may be sent to these islands for observing the transit of Venus in 1874. As it will be many years before more complete data can be procured and prepared, it may be useful to make the present results available to meteorologists and geographers generally.

From the registers kept in this region during the month of December, sets of observations were selected which would give the best daily mean values. They were generally six in number, though sometimes less. A day's observations were never divided, but all entered to the position of the ship at noon. The barometrical observations have been corrected for scale errors, for temperature, and reduced to sea-level. The temperatures have been corrected for any errors of thermometers exceeding half a degree. The specific gravity of the sea is given, corrected for errors of hydrometers and reduced to the standard temperature, 62° F. The directions of the wind have been corrected for variation of the compass.

The ships all sailed eastward through the region; those between 45° and 50° S. averaged latitude 46° 40′ S., and those between 50° and 53° S. averaged 51° 12′ S.

The years represented are the following :—

1855, 48 days; 1856, 6 days; 1857, 10 days; 1858, 6 days; 1859, 28 days; 1860, 4 days; 1861 6 days; 1870, 10 days : so that the observations embrace 118 days altogether.

For one particular day there are observations from 4 ships; for two other days from 3 ships; and for twenty-two of the days from 2 ships.

The distances apart of the ships were too great for satisfactory synoptic comparisons of weather, but their observations indicate the law of wind in relation to barometric pressure. In applying this law for the purpose of foretelling the wind's direction or its changes in high southern latitudes, the fall of barometer due to change of latitude should be taken into consideration.

The average heights of the barometer show the decrease of pressure for increase of latitude; but there is also considerable difference in respect to longitude. The barometer ranges higher about longitude 50° to 70° E., than to the eastward or westward. The temperature of the air is there also slightly higher. Between latitude 45° and 50° S., from longitude 40° to 80° E., the temperature of the sea increases from 39° to 45°; but between latitude 50° and 53° S., in the same longitudes, the sea is

---

* Note.—This paper has been prepared by Mr. R. Strachan, F.M.S.

remarkably uniform in temperature. The prevalent direction of the wind is NW. or WNW., and the force is usually high. In latitudes 45° to 50° S., longitudes 40° to 80° E., December appears to have on an average 5 very fine days, 11 fine, and 15 overcast : fog or mist occurs on 8 days ; rain, hail, or snow on 8 days ; and squalls on 4 days. In the same longitudes, but between 50° and 53° S. latitude, December averages 4 very fine days, 12 fine days, and 15 overcast ; fog or mist occurs on 6 days ; rain, hail, or snow on 10 ; squalls on 2 days. It is remarkable that with prevalent winds from NW. and WNW., the swell of the sea should predominate from W. and WSW.

The characters of the clouds, whenever noted, have been classified, and are as follows :

Lat. S.	Long. E.	Obs.	Cir.	Cir-c.	Cir-s.	Cum.	Cum-s.	Str.	Nim.
40 to 45	40 to 45	18	—	5	1	7	3	—	2
„	45 to 50	37	1	7	10	9	3	2	5
„	50 to 55	28	5	5	3	4	1	5	5
„	55 to 60	33	—	6	—	7	5	2	13
„	60 to 65	38	1	2	1	13	7	—	14
„	65 to 70	62	5	8	—	15	8	9	17
„	70 to 75	29	3	3	3	10	5	2	3
„	75 to 80	35	5	—	—	7	7	1	15
50 to 52½	40 to 45	14	—	3	1	6	3	1	—
„	45 to 50	10	—	3	1	3	3	—	—
„	50 to 55	20	—	—	—	8	4	—	8
„	55 to 60	17	—	2	—	1	6	—	8
„	60 to 65	13	—	—	1	2	9	1	—
„	65 to 70	21	—	5	3	—	8	—	5
„	70 to 75	21	—	3	2	2	9	—	5
„	75 to 80	18	—	—	1	—	12	—	5
53	40 to 60	18	—	—	2	—	10	6	—

From this it seems that the cirrus cloud was not seen south of latitude 50° S. ; and the stratus very seldom. Under nimbus, the few entries of "cumulonus" (FitzRoy) have been included.

In order to compare the mean dynamical motion of the winds with the mean barometrical pressure of the air, the resultant direction and force of the winds have been computed. This has been done in two independent computations. First, from the directions and mean forces by the Beaufort scale, as given in the table annexed ; and second, from the wind observations themselves, converting each estimated force into velocity in miles per hour, by the following table, which is an approximation to the various velocities corresponding to the several grades of Beaufort's scale, and has been drawn up from a consideration of the different estimates given by the several writers who have dealt with the subject.

Beaufort Scale.	Miles.	Beaufort Scale.	Miles.
0	2	7	42
1	5	8	50
2	10	9	60
3	15	0	70
4	20	11	80
5	27	12	90
6	35		

It will be seen from the results that the agreement is so close that practically it is a matter of indifference which method is adopted.

It also appears that the height of the barometer is related to the force of the wind ; the resultant wind being stronger where the barometer is lower, and *vice versâ*.

METEOROLOGICAL DATA FOR DECEMBER. Lat. 45° to 55° S. Long. 40° to 80° E.

Lat. S.	Long. E.	Barometer Mean	No. of Obs.	Air	No. of Obs.	Evaporation	No. of Obs.	Sea	No. of Obs.	No. of Obs.	Sp. g. of Sea Mean	No. of Obs.	b.	d.	o.	m.	t.	r.	s.	q.	lt.	Amount of Cloud.	Swell of the Sea noted from, on Days.
45 to 50	40 to 45	In. 29.386	26	39.3	26	38.1	21	39.0	22	3	1.0256	3	3	10	13	1	2	6	1	3		7.4	WSW. 3, Wb.N. 1.
„	45 to 50	29.391	38.	41.3	38	40.3	33	38.8	25	5	1.0256	5	5	12	21	5	2	2	4	5		7.4	WSW. 1, Wb.N. 1.
„	50 to 55	29.609	39	41.9	39	41.5	34	38.7	26	2	1.0248	2	13	15	14	10	1	8	4	6		5.9	WSW. 2, SSW. 2, Confused 1.
„	55 to 60	29.590	33	43.9	32	43.3	26	40.8	25	4	1.0254	4	7	15	35	17	5	16	1	3		8.1	W. 2, SSW. 2, ENE. 1, Confused 1.
„	60 to 65	29.538	44	45.8	44	44.6	39	42.9	37	6	1.0254	6	8	16	27	5	6	12	2	7		7.6	Wb.N. 2, WSW. 1, NE. 1, Confused 1.
„	65 to 70	29.576	72	43.5	70	42.8	65	41.8	49	5	1.0254	5		29	42	15	16	17	5	8		7.8	WNW. 2, WSW. 2, NE. 4, SSE. 1, Confused 2.
„	70 to 75	29.428	44	43.8	39	43.2	34	42.2	28	3	1.0255	3	10	19	18	3	9	14	2	10		6.9	WSW. 2, WNW. 1, Confused 2.
„	75 to 80	29.350	41	45.2	34	44.3	29	45.1	32	6	1.0244	6	6	25	14	2	5	16	3	11		7.4	Wb.N. 3, SW. 2, NNW. 1, Confused 1.
50 to 52½	40 to 45	29.128	18	37.4	8	36.8	12	37.5	9	1	1.0265	1	4	12	2	6			5	2		6.4	Wb.N 1.
„	45 to 50	29.263	23	36.6	17	36.1	23	37.5	14	4	1.0250	4	2	6	15	7		5	4	2		8.4	WSW. 1, NNW. 1.
„	50 to 55	29.215	34	37.7	18	37.1	28	37.4	13	5	1.0256	5	4	13	17	2		7	8	4		7.7	WNW. 1, WSW. 1.
„	55 to 60	29.394	28	37.6	12	36.5	16	37.3	3	3	1.0264	3	5	14	9	2	1		13	4		7.1	WNW.1, WSW.1, Nb.E.1.
„	60 to 65	29.602	22	38.1	11	36.3	16	37.6	8	3	1.0262	3	7	7	6	3			2			6.2	Wb.N. 2, WSW. 2.
„	65 to 70	29.356	23	38.1	14	38.1	11	37.4	11	3	1.0255	3	4	10	9	2	2	4	6	1		6.9	NW. 3, NE. 1.
„	70 to 75	29.590	28	38.5	17	37.1	22	37.4	11	3	1.0267	3	3	10	15	7	1	5	3	2		7.5	WNW. 2, WSW. 2.
„	75 to 80	29.570	34	37.9	17	36.7	17	37.0	12	5	1.0255	5			24	4	5	7	3			8.8	WSW. 2, NE. 2, Confused 1.
52½ to 55	40 to 45	28.961	6	36.3	6		6	36.3	6		1.0241			5	6	3	2	4				8.5	WNW. 1.
„	50 to 55	29.164	6	36.3	6	35.5	4	36.7	6		1.0240			3	1				2			7.0	NW. 1.
„	55 to 60	29.436	6	37.6	6	35.9	6	36.7	6	1	1.0240	1	2		2							6.3	WNW. 1.

WINDS IN DECEMBER. DIRECTION TRUE. MEAN FORCE (BY SCALE 0 TO 12).

N. O.*	N. F.*	NNE. O.	NNE. F.	NE. O.	NE. F.	ENE. O.	ENE. F.	E. O.	E. F.	ESE. O.	ESE. F.	SE. O.	SE. F.	SSE. O.	SSE. F.	S. O.	S. F.	SSW. O.	SSW. F.	SW. O.	SW. F.	WSW. O.	WSW. F.	W. O.	W. F.	WNW. O.	WNW. F.	NW. O.	NW. F.	NNW. O.	NNW. F.	CALM.	Variables O.	Variables F.	No. of Obs.	Resultants: Direction.	Resultants: For the Scale 0 to 12 (Miles.)	Resultants: Direction.	Resultants: Miles per Hour.	Lat. S.	Long E.
3	5·0	1	4·0	-	-	-	-	-	-	-	-	-	-	-	-	-	-	-	-	1	4·0	2	6·0	9	5·8	6	5·0	2	6·5	2	5·0	-	-	-	36	N 69° W	4·3=22	N 71° W	25·8	45° to 50°	40° to 45°
3	6·0	2	8·0	2	2·5	-	-	-	-	-	-	-	-	-	-	1	8·0	3	6·7	1	9·0	2	6·5	7	5·6	7	5·3	4	5·8	6	5·0	-	-	-	38	N 64 W	3·7=18	N 67 W	20·6	"	45 to 50
3	5·0	1	5·0	4	6·3	2	7·0	-	-	-	-	-	-	-	-	-	-	1	4·0	1	9·0	2	5·0	11	4·7	9	4·5	4	5·2	4	5·2	-	-	-	42	N 51 W	3·0=15	N 52 W	16·6	"	50 to 55
8	5·9	6	6·2	2	5·0	2	5·5	1	5·0	1	5·0	-	-	1	9·0	3	7·0	2	4·0	4	4·8	5	5·4	8	4·9	6	5·0	4	6·0	4	6·8	-	-	-	57	N 48 W	2·2=11	N 48 W	12·6	"	55 to 60
10	6·0	3	5·3	3	7·0	-	-	-	-	-	-	-	-	-	-	-	-	-	-	2	7·0	3	5·3	12	5·6	11	6·1	10	5·5	12	5·2	-	-	-	65	N 46 W	4·3=22	N 50 W	24·6	"	60 to 65
4	3·8	2	4·5	2	2·0	4	5·5	4	4·5	1	5·0	-	-	-	-	1	3·0	3	3·7	5	4·0	5	4·8	15	5·5	18	6·3	2	4·0	5	4·0	2	7	3·6	81	N 71 W	2·8=14	N 71 W	16·5	"	65 to 70
7	5·6	4	6·0	-	-	2	3·5	1	2·0	2	5·0	-	-	-	-	1	5·0	2	4·5	-	-	3	5·3	11	6·1	11	6·5	1	7·0	2	3·0	-	-	-	47	N 59 W	3·3=16	N 63 W	19·6	"	70 to 75
4	6·8	2	8·0	-	-	-	-	-	-	-	-	-	-	-	-	-	-	-	-	-	-	2	4·0	16	6·4	10	7·8	5	5·8	5	8·0	-	1	4·0	45	N 59 W	5·6=32	N 60 W	30·9	"	75 to 80
3	5·7	3	5·7	-	-	-	-	-	-	-	-	-	-	-	-	-	-	1	6·0	2	6·5	2	7·0	3	7·0	4	7·2	-	-	-	-	-	-	-	18	N 70 W	3·9=20	N 73 W	22·7	50 to 52½	40 to 45
10	6·3	2	6·0	-	-	-	-	-	-	-	-	-	-	-	-	-	-	-	-	2	5·0	2	3·0	1	3·0	1	3·0	1	5·0	2	6·5	2	-	-	23	N 15 W	4·0=20	N 15 W	22·9	"	45 to 50
4	5·2	-	-	1	7·0	1	3·0	-	-	-	-	-	-	-	-	-	-	-	-	3	4·0	3	6·3	16	5·9	5	6·2	-	-	1	5·0	-	-	-	34	N 79 W	4·3=22	N 80 W	25·0	"	50 to 55
2	5·5	1	7·0	1	7·0	-	-	-	-	1	5·0	1	6·0	-	-	1	6·0	2	6·0	5	4·6	5	6·0	3	5·3	2	5·0	2	5·5	2	5·5	-	-	-	28	S 84 W	2·5=13	S 82 W	15·4	"	55 to 60
3	4·0	1	5·0	-	-	-	-	-	-	-	-	-	-	-	-	-	-	-	-	-	-	1	5·0	4	4·5	6	6·2	5	6·4	2	5·0	-	-	-	22	N 53 W	4·6=24	N 54 W	26·1	"	60 to 65
1	7·0	3	7·3	-	-	-	-	-	-	-	-	-	-	-	-	1	5·0	1	6·0	2	5·5	2	5·0	2	5·0	5	6·2	3	6·3	3	5·7	-	-	-	23	N 56 W	3·6=18	N 57 W	22·6	"	65 to 70
1	6·0	2	6·0	-	-	-	-	-	-	-	-	-	-	-	-	-	-	2	4·5	4	4·8	2	5·0	5	7·4	4	6·2	5	5·8	3	5·0	-	-	-	28	N 72 W	4·0=20	N 73 W	24·0	"	70 to 75
3	7·0	2	6·5	1	1·0	2	2·0	1	3·0	2	3·0	2	3·0	1	3·0	1	2·0	-	-	1	5·0	2	5·5	3	4·0	2	4·0	6	4·0	4	3·3	1	-	-	34	N 33 W	1·8= 9	N 36 W	11·1	"	75 to 80
-	-	-	-	-	-	-	-	-	-	-	-	-	-	-	-	-	-	-	-	-	-	-	-	1	8·0	2	8·0	2	8·0	1	8·0	-	-	-	6	N 50 W	5·6=32	N 54 W	33·3	52½ to 55	40 to 45
-	-	-	-	-	-	-	-	-	-	-	-	-	-	-	-	-	-	-	-	-	-	-	-	-	-	1	7·0	3	6·7	2	6·0	-	-	-	6					"	50 to 55
-	-	-	-	-	-	-	-	-	-	-	-	-	-	-	-	-	-	-	-	-	-	-	-	1	4·0	1	4·0	2	3·5	2	8·0	-	-	-	6					"	55 to 60

OBSERVATIONS OF WIND DIRECTION, FROM MAURY'S PILOT CHARTS.

-	-	2	-	-	-	-	-	-	-	-	-	-	-	-	-	-	-	-	-	-	-	12	-	1	-	1	-	-	-	12	-	1	-	-	39	—	—	—	—	45 to 50	40 to 45
1	-	6	-	-	-	-	-	-	-	2	-	1	-	3	-	-	-	11	-	-	-	6	-	6	-	28	-	7	-	5	-	-	-	-	76	—	—	—	—	"	45 to 50
5	-	2	-	-	-	2	-	-	-	-	-	-	-	-	-	-	-	-	-	-	-	-	-	-	-	20	-	3	-	18	-	-	-	-	49	—	—	—	—	"	50 to 55
-	-	-	-	-	-	-	-	-	-	-	-	-	-	-	-	-	-	-	-	-	-	3	-	-	-	-	-	3	-	3	-	-	-	-	9	—	—	—	—	"	55 to 60
-	-	-	-	-	-	-	-	-	-	-	-	-	-	-	-	-	-	1	-	-	-	-	-	-	-	3	-	4	-	-	-	-	-	-	8	—	—	—	—	"	60 to 65
-	-	1	-	-	-	2	-	-	-	-	-	-	-	-	-	-	-	11	-	-	-	-	-	4	-	6	-	2	-	4	-	-	-	-	29	—	—	—	—	"	65 to 70

* The headings O. and F. are for "Number of Observations" and "Mean Force."

NOTES FROM THE OBSERVER'S REGISTER.

Date.			S. Lat.		E. Long.		Remarks.
Y.	M.	D.	°	′	°	′	
1855	12	21	48	20	43	30	Passed a large iceberg, afterwards a smaller one; several icebergs, and stormy petrels.
1859	12	5	46	10	44	25	Great numbers of birds; several patches of seaweed.
1855	12	11	49	30	49	0	Passed two small bergs.
„	12	6	48	55	48	56	An iceberg about a mile in diameter.
1859	12	1	46	21	49	50	A large quantity of kelp; several seals and penguins.
„	12	6	46	0	47	0	A large iceberg.
1855	12	23	48	30	54	0	A large iceberg, about 200 feet high, and about a mile long. Several icebirds, Cape pigeons, petrels, and albatrosses; afterwards two large ice islands.
„	12	13	48	14	50	50	Two large bergs.
„	12	3	46	0	52	0	Water discoloured, a quantity of seaweed and ice birds.
„	12	4	45	42	55	41	A quantity of seaweed and a number of ice birds.
1859	12	3	46	0	58	0	Water more greenish than usual; seaweed.
1858	12	27	45	30	58	50	A land bird; several ice birds; a few albatrosses.
1859	12	29	45	11	55	8	Large quantities of seaweed.
1855	12	24	48	10	60	44	Water a lighter colour.
1859	12	4	45	46	63	31	A quantity of kelp.
1859	12	31	45	26	63	24	A little seaweed.
1858	12	30	45	18	68	27	Large quantities of seaweed.
1859	12	5	45	43	66	42	A quantity of penguins; shoal of porpoises going southward; large quantities of seaweed.
„	12	6	47	6	68	31	Shoal of porpoises going eastward; water more greenish than usual.
„	12	7	47	40	69	29	Plenty of albatrosses and penguins; four whales. A great many albatrosses.
1855	12	27	47	40	74	22	A small piece of ice.
„	12	17	47	14	70	53	Water very much discoloured; a quantity of birds about.
„	12	10	47	26	73	12	The water very much discoloured.
„	12	11	48	30	72	0	Seaweed.
1861	12	9	45	54	76	26	A few birds.
1855	12	18	47	17	76	21	Water discoloured; a few birds.
„	12	12	48	30	77	31	Some seaweed.
1859	12	11	46	12	76	0	Several patches of seaweed.
1855	12	24	51	0	41	0	Four icebergs.

NOTES FROM THE OBSERVER'S REGISTER—*continued.*

Date.			S. Lat.		E. Long.		Remarks.
Y.	M.	D.	°	′	°	′	
1856	12	15	51	0	45	0	An iceberg about 200 feet high ; also a piece of loose ice.
1855	12	25	51	0	46	0	A large iceberg and a quantity of small pieces.
1856	12	16	51	12	51	8	A few petrels.
1855	12	13	50	30	59	0	A large iceberg, distant four miles, water 36°, afterwards rose to 39°.
1856	12	18	51	38	62	20	Sperm whale.
1855	12	29	51	40	62	45	Two icebergs and four large pieces, each piece about half the size of the ship ; the sea breaking over them.
1857	12	15	50	20	68	8	A number of petrels.
1855	12	30	51	40	68	0	Sea high ; height of waves 30 feet, distance between each 170 feet, rate about 1½ knots ; sea quite green.
„	12	15	50	4	72	4	Water greenish.
„	12	31	51	30	73	50	Sea quite green.
1856	12	19	51	38	70	16	Water discoloured a little. A few snow birds, Cape pigeons and sooty albatrosses.
1857	12	16	50	15	73	43	Sea mud-like colour.
„	12	17	50	55	75	35	Water a dull green ; a number of sea birds.
„	12	18	50	47	77	10	Two whales; sea dark blue.
„	12	19	50	59	77	52	A flock of sea ducks.
1856	12	20	51	38	77	24	Water much discoloured ; examined some of it with a microscope ; some few animalcules, but not enough to colour the water. Birds about, particularly *diomedea familiaris* (*sic*) and Cape pigeons.
1857	12	17	50	50	76	20	At midnight, aurora in SE., spreading its rays sometimes as high as the zenith.
1870	12	26	53	0	56	0	A small iceberg.

LONDON :
Printed by GEORGE E. EYRE and WILLIAM SPOTTISWOODE,
Printers to the Queen's most Excellent Majesty.
For Her Majesty's Stationery Office.
[3873.—600.—12/79.]